Telecom Nation
Telecommunications, Computers, and Governments in Canada

Governments around the world have found the massive expansion of telecommunications systems and the breathless pace of innovation too important to be left to the market alone. In Canada, telecommunications became an important object of public policy. *Telecom Nation* focuses on how governments and regulatory agencies handled the communications revolution in the three critical decades after the Second World War.

Laurence Mussio examines how federal and provincial public policy tried to keep pace with the diffusion of telecommunications, consumer demand, and a rising tide of technological innovation. Telecommunications regulation struggled to maintain a balance between producer and consumer in an increasingly complex field and policy makers were compelled to defend the national interest in international telecommunications arrangements by making far-reaching decisions about transcontinental microwave systems and satellites. By the late 1960s national policy makers had embraced the arrival of the computer – especially once it began to be wired into Canada's communications infrastructure. *Telecom Nation* explores the impact of the computer on government policy and the first attempts to build a "national computer utility" – the beginnings of the Internet – twenty-five years before it became a reality.

Based primarily on the rich and largely untapped sources at the national Archives of Canada, cabinet records, provincial archives, and private sector repositories, *Telecom Nation* provides an essential background to contemporary public policy issues by examining how governments reconciled technological change, private enterprise, consumer demand, and the public good in communications. It will be required reading for students and specialists interested in telecommunications, public policy, and technological change.

LAURENCE B. MUSSIO is a senior communications consultant in the financial services, communications, and public sectors.

Telecom Nation

Telecommunications, Computers, and Governments in Canada

LAURENCE B. MUSSIO

McGill-Queen's University Press
Montreal & Kingston · London · Ithaca

© McGill-Queen's University Press 2001
ISBN 0-7735-2175-5

Legal deposit second quarter 2001
Bibliothèque nationale du Québec

Printed in Canada on acid-free paper

This book has been published with the help of a grant
from the Humanities and Social Sciences Federation of
Canada, using funds provided by the Social Sciences
and Humanities Research Council of Canada.

McGill-Queen's University Press acknowledges the
financial support of the Government of Canada through
the Book Publishing Industry Development Program
(BPIDP) for its activities. It also acknowledges the
support of the Canada Council for the Arts for its
publishing program.

Canadian Cataloguing in Publication Data

Mussio, Laurence B.
    Telecom nation : telecommunications, computers, and
governments in Canada
    Includes bibliographical references and index.
    ISBN 0-7735-2175-5
    1. Telecommunication policy – Canada. 2. Computer
networks – Government policy – Canada. I. Title.
HE7815.M88 2001   384'.0971   C00-901244-3

Typeset in Palatino 10/12
by Caractéra inc., Quebec City

For Flavia

Contents

Acknowledgments

Telecom Nation was first conceived as a project when I was a member of the Graduate Program in History at York University in the mid-1990s. The department's rigorous standards and its encouragement of young scholars made it a perfect environment for research and writing. I have been fortunate to have learned from some of the finest practitioners of the historian's craft in my field. Funding for this book came from the Social Sciences and Humanities Research Council of Canada and the Walter L. Gordon Foundation, whose assistance was vital and is gratefully acknowledged. The Humanities and Social Sciences Research Council of Canada's Aid to Scholarly Publications Program lived up to its name and made the publication of this book a reality. These programs ensure that scholarship can flourish in Canada and contribute to a broader and deeper understanding of our national experience.

I owe a special thanks to the archivists and staff of the National Archives of Canada and the National Library of Canada, the Privy Council Office of Canada, the Archives of Ontario, and numerous public and private repositories who met my prolific requests for documents with patience and professionalism. I would especially like to thank Stephanie Sykes, former archivist at Bell Canada, and Lorraine Croxen, currently of Bell Canada Archives in Montreal, for their gracious assistance at the beginning and end of this project.

This study took shape under the discerning eye of H.V. Nelles. His exacting supervision of my work combined dispassionate criticism, a passion for inquiry, and a seemingly inexhaustible intellectual power supply. He has taught me far more about history than a simple acknowledgment can convey. My thanks are more than formal. I also extend my thanks to Drs Christopher Armstrong and Frederick Fletcher of York University, and Dr Duncan McDowall of Carleton University for commenting on earlier drafts of the manuscript. The anonymous press readers were especially helpful in providing a valuable perspective on the manuscript. Their comments prompted me to pose new questions of my evidence.

Presses can be arenas of spiritual trial and suspense for authors – places where manuscripts are transformed, grammatical realities confronted, arguments challenged, and value adjudicated. Relations between authors and publishers can therefore either be fraught with tension, or flow peacefully into print. At McGill-Queen's University Press, I have been fortunate to have enjoyed the latter experience for not one, but two major manuscripts. I extend special thanks to Mr Philip J. Cercone of the Press for his unfailingly kind encouragement and invaluable assistance in the publication of this study. Through his helmsmanship of the Press, Mr Cercone has made an important and original contribution to Canadian and international scholarship. Joan McGilvray supervised the many details of the manuscript as editor. Diane Mew copy-edited the manuscript, sharpened both prose and argument, and provided many hours of good conversation besides. I thank them both.

I would like to express my gratitude to my mother, father, and sister Louise, who sustained me throughout my scholarly pursuits and never once asked when it would be done. I also extend a special thanks to my uncle, Monsignor Bruno F. Pighin, whose high scholarship and public spirit have been both model and example to me. My friends and family provided diversions from long days in archives and writing, and one, Christopher Rupar, provided all of that as well as something more concrete (in more ways than one) – an Ottawa couch during my long and frequent visits to the National Archives.

I would especially like to thank my wife Flavia for her splendid support, encouragement, and utter lack of interest in the subject. This formula has been successful through concurrent manuscripts, and has kept me well grounded and able to focus on the wider world. To her, I dedicate this work. Together, all of the people mentioned above have made the writing of this book a rewarding experience, and its publication a reality.

Toronto, Canada
August 2000

"WE'RE ON OUR WAY!"

... With the biggest construction programme in our history. It will provide the rural telephone equipment which could not be installed during the war.

RURAL TELEPHONE HIGHLIGHTS FOR 1946

TELEPHONES ... Over 12,000 new rural telephones were added, bringing the Company total to 100,000.

LINES ... Nearly 1,100 new rural telephone lines were built to provide for still more telephones with fewer parties on each line.

SWITCHBOARDS ... Six exchanges and 2,000 telephones were changed from magnetic (crank) operation to the modern "common battery" system as used in many large cities. It is planned to change over 20 more exchanges and some 6,000 telephones in 1947.

CALLS ... Efficiently and courteously, more operators completed more local and long distance calls than ever before.

Our $5,000,000 rural construction programme is being pushed at top speed so that you may continue to have the best telephone service at the lowest cost.

THE BELL TELEPHONE COMPANY OF CANADA

A Bell Telephone advertisement illustrating and promoting expansion. Courtesy Bell Canada Historical Collection.

Telephones galore. Bell Telephone Advertisements, *circa*
1949, illustrating Bell's remarkable expansion in the 1940s.
By the end of that decade, half a million phones were
added to the Bell system. Courtesy Bell Canada Historical
Collection.

This advertisement warns that massive expansion took "lots of work and lots of money" – a job that only a financially healthy company could perform. Bell Telephone Advertisements, 1950. Courtesy Bell Canada Historical Collection.

Bell's in-house organ, *Telephone News,* announces the new microwave radio-relay "skyway" linking the Canadian Broadcasting Corporation's television stations in Toronto, Ottawa, and Montreal, May 1953. Courtesy Bell Canada Historical Collection.

YOUR VOICE
TAKES THE
HIGH ROAD...

Most maps don't show the mining sites, fishing lodges and camps where people live and work or play in Northwestern Ontario. But your voice, transmitted from a high power similar to the one shown, located at Kashabowie, will reach them — and they can call to anywhere in Canada or the world via Bell's Fringe Radio Telephone Service.

This installation is one of many in Northwestern Ontario built by The Bell Telephone Company to bring service to remote locations. Bell people continually survey the territory they serve, and as soon as they discover a need for phone service, something is done about it: it may be Fringe Radio Service in distant parts of Northwestern Ontario, or new buildings, exchanges and equipment in towns and cities.

Throughout Northwestern Ontario, Bell serves some 18,000 telephones directly through 41 exchanges and adds to its stake in the area (now over 25 million dollars) at the rate of $2 million every year. And Bell's material resources are matched by its people, many of them actually from the region, who strive to give the best communications service possible, in keeping with their own and their Company's faith in the bright future of this growing area.

THE **BELL** TELEPHONE
COMPANY OF CANADA

Built, managed and owned by Canadians

Taking the high road: Bell advertises its high-power radio transmitter in northwestern Ontario to bring service to remote areas of the province. Courtesy Bell Canada Historical Collection.

Construction on the Bell Network. Above: Installation 1940s. Below: Laying cable, 1950s. Courtesy Bell Canada Historical Collection.

TELEPHONE

NEWS

To give you a better understanding of our business and the services we provide.

FIRST IN CANADA

Direct Distance Dialing To Start Here July 8th

Intricate accounting equipment like this will keep track of the out-of-town calls you dial yourself. By means of a perforated tape, this apparatus accurately records full details of the call so that it can be billed to your account.

A fast, easy way of dialing your own out-of-town calls — known as Direct Distance Dialing — will be introduced in Windsor, La Salle and Tecumseh beginning Sunday, July 8th.

From your home or office telephone, you'll be able to dial to any telephone in Detroit and more than 60 other Michigan cities and towns. You'll use this speedy "dial-it-yourself" system on all "station-to-station" calls (those where you speak to anyone who answers). Other calls — such as person-to-person calls and those made from public 'phones — will be made through the Long Distance operator as usual.

A special calling guide to insert in your telephone book, plus a Blue Book for Direct Distance Dialing, is being mailed to everyone shortly before July 8th. These contain a complete list of the places you can dial direct and full instructions on how to do it.

This is the first area in Canada to get Direct Distance Dialing. In the next few years, as the new system is extended, you'll be able to dial direct to telephones in most parts of Canada and the United States.

Windsor, June 1956

It's Important to Remember...

You'll find full instructions for Direct Distance Dialing in the calling guide and special Blue Book being mailed to everyone. But these points are worthy of re-emphasis:

1. *Keep an up-to-date list of out-of-town numbers you call.* You'll be able to put your calls through faster when you know the number you want.

2. *Be sure to dial carefully.* If you do dial a wrong number, try to find out the place and number you reached. Then dial OPERATOR right away and tell her about it, so you won't be charged for the call.

In 1956, Bell introduced Direct Distance Dialing in Windsor, Ontario, 1956. Courtesy Bell Canada Historical Collection.

Like many large companies, Bell installed a Honeywell 400 computer in 1963 to handle Treasury stock transfers, Yellow Pages advertising contracts, and traffic dial assigning. Courtesy Bell Canada Historical Collection.

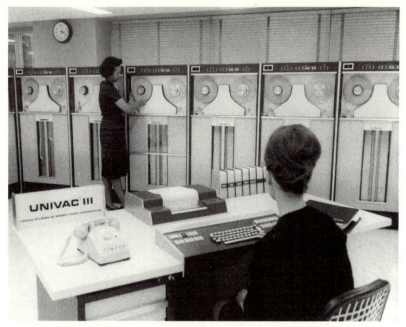

The Univac III Computer, installed by Bell Telephone for billing and collection purposes in 1964, was retired from service in 1982. Courtesy Bell Canada Historical Collection.

Bell Canada Satellite Earth Station, Lac-Bouchette, Quebec 1968. Courtesy Bell Canada Historical Collection.

Advertisement for Dataroute, the world's first country-wide digital data transmission service available to the public. Courtesy Bell Canada Historical Collection.

Telecom Nation

Introduction

For governments around the world in the twentieth century, the massive expansion of telecommunications systems was thought to be too important, too critical for the life of a country, to be left to the market alone. A technology that reached into every facet of ordinary life needed to be carefully channelled. Even more than the telegraph before it, the telephone carried potentially enormous social and political consequences. Telecommunications also spawned institutional behemoths which commanded huge territorial monopolies and wielded immense economic power. Whether managed by businessmen or bureaucrats, the organizations created to harness these new forms of communications were, in every sense of the word, public: they depended upon public law, used public property, provided public services, and so became public utilities.

Like other massive technological transformations in transportation, light, heat, and power, the rapid spread of new communications systems forced governments of all stripes to respond. In continental Europe and the United Kingdom, governments quickly asserted control and ownership of national telecommunications networks. In the United States, private companies were allowed to manage systems and provide services, under the supervision of both state and federal regulation to police rates and ensure equitable provision of service. In Canada, both US and European models were adopted, and domestic hybrids combining both also flourished. A range of regulatory instruments, including state commissions, regulatory tribunals, government departments, and government monopolies were put into place across the country. The political and social compromises reached over economic organization, technology, and public choice were thus cast into a series of institutional arrangements devised to protect both producers and consumers and destined to last well after the Second World War.

As an economic and technical development, telecommunications in the first half of the twentieth century featured one major innovation (telephony) and a steady pace of diffusion. Governments seeking to balance competing interests, ensure national sovereignty, or protect their own local power in the growth of telecommunications could therefore manage to do so comfortably with the regulatory instruments they had created over time. Stability was easily maintained when the players and the technology did not change all that much.

In Canada, this relationship between public power and communications became a vital one, and for several reasons. Vast distances within the country made efficient transportation and communications necessary for political and economic survival. The urbanization and industrialization of Canadian life cemented the country's reliance on technological systems. Canadians would become notorious for having more telephones per capita than almost anywhere in the world, and making loquacious use of them as well. Later, some of the best Canadian communications utilities put what they had learned into developing something of an international specialization in the sector. Government regulation before 1945 confined itself to seeing that telecom monopolies made a reasonable profit, consumers had reliable telecommunications at a more or less reasonable cost, and broader objectives such as ensuring system development or furthering regional economic objectives could be achieved. As long as those conditions were met, a broad consensus reigned.

A remarkable series of technological developments after the Second World War produced an explosion in communications which eventually upset the balance between public power and technology and had serious consequences for the ability of public policy to influence technological development. Transcontinental microwave networks and newly laid ocean cables in the 1950s, and satellite technology and the beginnings of computer communications in the 1960s, were only the most glittering in the array of new telecom technologies. Small but continuous improvements in telephony, and especially long-distance calling, made telecommunications technology universal, accessible, and indispensable. The growing power and reach of those technological systems slowly but relentlessly undermined the old arrangements. The companies that provided these services in Canada, most notably the Bell Telephone Company of Canada, became major institutional players in Canadian economic life. When new but complementary systems based upon electronic technology arrived, an American company eventually asserted its supremacy: International Business Machines (IBM). Both companies in their respective national and international contexts were (and still are) technological giants. Although each operated in markedly different contexts, both can be said to have grown up in the golden age of monopoly whether it was protected as Bell was, or unprotected as IBM was. Bell enjoyed a pre-eminence in the two decades after the Second World War unsurpassed since the company's inception in 1880. IBM positioned itself instead atop a massive technological tide created in the United States by both government and business and extended its technological dominion systematically around the world. Although IBM did not possess a monopoly, the company's position was based on an almost monopoly-like control of the marketplace in computers.

These technological developments posed extraordinary challenges for both public policy and regulation in Canada. At the end of the Second World War, Canada was on the threshold of a technological leap in telecommunications. How did Canadian institutions manage these massive technological changes with, essentially the economic and regulatory models inherited from the three decades before 1945? How did public policy and regulation cope with these changes when they were actually happening?

Governments in this period were compelled to respond to these challenges on many fronts: through seeking to protect Canada's interests in international communications, or by developing a domestic telecommunications policy. By the 1960s, the focus on telecommunications also embraced the most momentous econo-technical development of the half-century – the computer – especially once it began to be wired into Canada's communications infrastructure. While policy-makers struggled to come to terms with potentially enormous changes, regulators laboured to balance the imperatives of development with popular pressure over rates and service. These changes not only threatened to transform markets, but the ability of governments to respond to them. When governments did respond, they did so in often contradictory ways. Canada's encounter with this postwar transformation in communications, and the consequences of the interaction between public power and technological change, is the subject of this book.

The subjects discussed here have inspired both scholarly and popular attention from an astonishing range of writers – from economists to political scientists, to futurists, communications theorists, and even the occasional millennial prophet. Their concerns have ranged from analysing the behaviour of telecom firms under regulation, finding out what went "wrong," or exposing tyrannical monopolistic behaviour. Others have focused on the possibilities of revolutionary technology to transform contemporary life. Recent literature on the modern regulatory state has provided some perspective on telecommunications policy both in the contemporary North American and in Canadian contexts.[1] By offering this analysis of the historical relationship between Canadian public policy and communications technology, I hope to fill a conspicuous gap in the literature and add to the historical understanding of a critical sector in Canadian economic life.

To capture the breadth of technological transformation experienced in Canada after 1945, I have also examined public policy responses to the arrival of the computer in Canada. At first glance, the fusion of computers and telecommunications in this study may seem somehow out of place. Computer technology was not subject to regulatory control of governments, at least to the extent telecoms were. The development of the Canadian telecommunications sector, moreover, could be considered in many ways a story in which both the private and public sectors were key players, and had an important influence over outcomes. By contrast, public policy in the computer/communications sector was diffuse, and policy-makers watched from the sidelines.

Nevertheless, from a contemporary point of view there are plenty of reasons for studying computers and telecommunications together. The marriage between the two has produced the information highway, and wired Canadians into a technological transformation of the first order. My broader objective is therefore to show how two separate technological tracks emerged, then how they began to converge. The emergence of computers and the creation of an embryonic information sector posed special challenges to public policy. Policy-makers of the 1960s witnessed these developments and were seized with the massive potential that a union of telecoms and computers would bring. Their vision led them to try to build the first information highway, twenty-five years before it became a reality in Canada. How policy-makers responded to those challenges with the tools they had, and the assumptions they held, can shine a light on how governments dealt with major technological change as it was happening.[2]

The reconstruction that emerges from these pages has been built from private correspondence, memoranda, cabinet minutes, private papers, and transcripts of regulatory hearings drawn from a range of public, agency and private archives across the country. In many instances, the archival evidence presented here is being discussed for the first time. Sometimes this evidence confirms what many suspected; in many other instances, this original research has unearthed submerged parts of familiar stories.

From this archival base, I hope that this book will shed light on how the Canadian state navigated through an era of major technological transformation. That transformation left nothing untouched – from the mundane but important business of regulating telephone rates to the high policy of designing a national telecommunications strategy, or even a national computer utility. At the very least, this book will enable comparisons to be made with the experience of other countries in telecommunications and high technology policy.

Telecom Nation represents an historical approach to its subject. This has implications for the way telecommunications, computers, and government policy are presented. Theoretical considerations, for example, are largely absent from most of the discussions in this book, except where they are useful in explaining particular actions. This is the view from national, provincial, and local archives and agencies, as well as from a wealth of private-sector repositories. Correspondence, memoranda, hearing and tribunal transcripts, briefs, private papers, newspaper accounts, all form the basis for this study.

I have also drawn upon a rich source of interdisciplinary literature in telecommunications in regulatory economics, political science, and law. This book has built upon those insights without necessarily imposing the theoretical constructs that often go with them. In this book, the interplay of will, circumstance, and personality combine with the great waves of technological development and the rules of the marketplace. It is my hope that this approach will offer a more organic picture of the development of government policy in telecommunications and computers in Canada.

This study ends in 1975, for two reasons. First, it was in 1975 that the major regulatory power over telecommunications was transferred from the Canadian Transport Commission to the Canadian Radio-Television and Telecommunications Commission (CRTC). The CRTC's jurisdiction over telecommunications thus ended one era of regulation and began another. Secondly, this book is a work of historical scholarship. The quality, nature, and extent of archival sources in the post-1975 period (especially in the National Archives), not to mention access restrictions, make the extension of such a definitive study almost impossible at this time. Further, since this is an historical study, I have made certain to keep a space of at least two decades from my subject. The past is usually better judged from a distance.

For reasons of space, this study will only consider questions arising out of "hard wire" problems: carriage, not content, is examined.

Consequently, the history of broadcasting and cable is not substantively discussed. The history of Canadian broadcasting is a relatively well-covered subject, and involves a different set of historical actors and circumstances. Neither is it a corporate history of the industries involved in these activities; rather, it examines the attempts of the state to defend its interests while mediating between the competing interests of others. Readers accustomed to hearing of the ringing historical success of telecommunications in Canada, the strength of its companies, the glamour of its BCEs and Nortels, may be surprised at the focus and the conclusions of this study. There are reasons for this. *Telecom Nation* concentrates upon the role of government – in both regulation and policy – on what government has done, and what it has failed to do. The success of Canadian telecommunications does not mean, for example, that relationships between governments and companies stayed the same, or that the balance of power between the two did not shift. Sometimes the system worked despite of, not because of, regulation.

For those unfamiliar with the pattern of industrial organization of telecommunications in Canada, some information needs to be provided first. Responsibility for telecommunications in Canada is divided by constitution and evolution. Although several provinces either regulate or operate systems in their own territory, the great bulk of the Canadian telecommunications is regulated at the national level. Three principal utilities were regulated by the federal government: the Bell Telephone Company of Canada, which operated in Ontario, Quebec, and the Northwest Territories; British Columbia Telephone which operated in British Columbia; and CN/CP Telegraphs (later Telecommunications), which operated nationally. Overseas telecommunications links were controlled by the Canadian Overseas Telecommunications Corporation (COTC, later Teleglobe Canada) established by federal statute in 1949. Together, these operators represented approximately three-quarters of the Canadian system. The federal government retained constitutional jurisdiction over "lines of steam or other Ships, Railways, Canals, Telegraphs" as well as other works which extended across provincial boundaries. Federal responsibility over the regulation and control of radio communication was reaffirmed in 1932 by the Judicial Committee of the Privy Council which ruled that the Canadian Parliament had exclusive legislative power to regulate and control radio communication in Canada.[3] Communications were considered a matter of national importance requiring a national scope.

At the federal level, the Board of Railway Commissioners (BRC) was given the protean responsibility of overseeing both national transportation and communications. In 1906 the BRC acquired jurisdiction over federally chartered telephone and telegraph companies. In 1938 the BRC became the Board of Transport Commissioners for Canada (BTC). In 1967 the name was changed once again, to the Canadian Transport Commission (CTC). But the tasks were the same: to provide a good business climate for the utilities it regulated while balancing the interests of subscribers. The agency's most important task was mediating between competing interests and maintaining the legitimacy of the process. Given the size of the utilities under BRC/BTC/CTC control, federal regulatory decisions carried considerable scope and weight.

One of the most striking features about Canada's experience with telecommunications is the way that Canadian institutions and ways of thinking made their mark on the technology, and not the other way around. Deeply held convictions about the nature of federalism, sovereignty, community property, economic development, and technological innovation and change on the northern half of the continent helped to determine the Canadian approach. In other words, telecommunications technology not only influenced Canadian economic development; Canadian markets and Canadian political arrangements themselves shaped how this technology would be dealt with. The institutional memory of government and historical patterns of thinking shaped how new technologies and their significance were perceived. Past transformations in transportation, communications, and banking, for example, deeply influenced how people thought telecommunications could be harnessed to national objectives, or even what was considered to be at stake.

Part One focuses upon the most important developments in telecommunications policy and regulation between the end of the Second World War and the 1960s. Chapter 1 is set in the contested environment of federal telecommunications regulation where commissioners, companies, consumers, and occasionally the federal cabinet struggled over questions of cost, system development, and monopoly power. Chapter 2 examines the Canadian government's responses to the rise of a new global telecommunications order after the Second World War that flowed as much from technological advance as from the new political arrangements of the postwar world. The response Canadian policy-makers to developments in computer technology in chapter 3 provides us with an opportunity to examine the possibilities, as well as the limitations, of Canadian public policy in the face of an exceptional technological development. By the 1960s, the arrival of both microwaves and satellites drove the federal government into coming up with a national telecommunications policy. That story is the subject of chapter 4.

In Part Two the focus turns to the late 1960s to mid-1970s and the attempts to produce a national telecommunications policy on the one hand, and an increasingly paralysing conflict over regulation on the other. Chapter 5 analyses the slow collapse of federal telecommunications regulation. During this period, federal telecommunications regulation was plunged into crisis, unable to cope with growing demands for broader public participation and ultimately unable to sustain its legitimacy. At the same time, the federal government launched a concerted effort to create a strong national communications policy in the face of major technological change. As chapter 6 shows, internal weakness within the federal government and federal-provincial conflict sealed the fate of that attempt. In chapter 7, Ottawa's ambitious plans for fusing computers and communications into a national computer utility as the first "information highway" in the early 1970s provides us with a fascinating case study of what can happen (and what often doesn't) when politics and technology converge. Canada's computer communications policy, moreover, gives us an opportunity to assess the what policy-makers thought they saw in the arrival of new technologies, and what they actually did see – which often turned out to be two very different things. It will strike anybody familiar with the contemporary configuration of the Internet as a strange episode from a distant place. Finally, the conclusion analyses the postwar Canadian experience in telecommunications and its implications for the relationship between public power and technological development in the telecommunications and high technology sectors.

PART ONE
The Technological Imperative

1 From Golden Age to Iron Cage: Telecommunications Regulation and the Board of Transport Commissioners for Canada, 1945–1966

The emergence and remarkable technological diffusion of electric power, telephony, and mass transit across the late nineteenth- and early twentieth-century urban landscape had created massive monopolies. Civic populists and municipalities demanded that these monopolies be responsible to the community. What materialized from the ensuing conflict between these opposing forces was a spectrum of regulation, from the independent regulatory commission to outright public ownership. Regulation not only attempted to achieve a balance between producer and consumer, it also determined the pace and speed of technological development in Canada.[1]

For the Bell Telephone Company, this arrangement meant something very specific: it had to submit its operations to scrutiny by the federal Board of Railway Commissioners (BRC) in 1906, which determined the rates Bell could charge for its services.[2] Between Bell's first application for a rate increase in 1907 and its last interwar application in 1927, the board was a cooperative agent in respecting managerial prerogatives about rates and system expansion. The board was also helpful in shielding the company from political pressure. By subordinating regulatory discourse into the narrowly defined statutory parameters of the Railway Act, a certain balance could be struck between both the regulators and regulated, and between company and consumer. The political and economic bargains struck between public and private power provided its foundation; regulatory practice gave it shape and substance. Between its creation and the Second World War, the BRC (and from 1938, the

Board of Transport Commissioners) produced a distinctive style of regulation in telecommunications.[3]

The regulation of telecommunications after the Second World War has received little attention, save the occasional comment that regulation was limited to policing rates. While this is true, it is only the beginning of a much more significant and complex story. The two following decades were critical ones for the architecture of Canadian telecommunications. The explosion in demand for telecommunications on the heels of Canadian postwar growth positioned Bell atop an expansion curve that proved to be both durable and profitable. Telecommunications was assuming an indispensable importance in Canadian economic life; the agency and the companies it regulated could both boast of a rapid diffusion of technological benefits; and its status as an absolute social and economic good was unquestioned. And if technology was faith, Bell was its church. This ethos of expansion was forged in the full flush of Canadian postwar economic growth, and based upon a solid foundation of consumer demand. This consensus fuelled maximum system development and held out the promise of ever-greater technological benefits to its subscribers. Mass consumption of telecoms could even be shaped to progressive economic and political ends.[4]

Technological diffusion and falling costs are certified by the statistics. As tables 1 and 2 show, in both absolute and relative terms, between 1918 and 1975 long-distance prices between Canadian cities have fallen dramatically. As the next table dramatically illustrates (table 3), the number of telephones had gone from over 2 million in 1946 to 13.1 million in 1975, while the number of telephone calls has skyrocketed from 3.5 billion to 21.2 billion.

Demand and expansion, however, had its effects in the balance between public and private power. A technological imperative tipped the balance of regulatory power between agency and utility in favour of Bell. Bell's expansion was partly self-financing; but when matters did force a public hearing, the BTC could be counted on to produce a decision sympathetic to the company. The result was a serious erosion of the public power in shaping the development of telecommunications. It should be clear that this was an ethos widely shared among regulators, company executives, and many Canadians besides. It was an outcome of the dynamics of the process, a consensus that produced consistent technological benefits.

But this ethic of development was often contested, and there were serious challenges to regulation. Opposition to rate increases came from those who were reluctant to pay higher rates for telephone service, and also from groups who were unconvinced that the benefits of network expansion and technology were enough to cancel out more grave concerns about monopolistic extravagance. Familiar exasperations about Bell's efficiency, excess, and imperious style of management merged with new concerns about the phenomenal growth of the utility. Together, they stimulated political conflict. The opposition largely fell to municipalities, the original grantors of telephone franchises, and as well established participants in the regulatory process, to speak for the consumer and to argue for restraint in Bell operations and expansion.[5] It was hard to argue with success.

In that context, then, regulation had to navigate between residential and corporate interests, and between company and consumer. At the same time, the BTC had to ensure that the needs of the telecommunications network were accounted for as well. As telecommunications development both picked up speed and became more complex, telecom regulation sought to keep up. As the stakes got bigger and bigger, federal regulation provided an exasperating demonstration of its inability to reconcile conflicting visions of the country's technological development. As a result, public power in telecoms suffered. The course of that regulation is the subject of this chapter.

A REGULATORY CULTURE

Between 1906 and the 1930s, the commissioners of the BRC and BTC created a distinctive regulatory culture that set it apart in dealing with Canadian telecommunications development. In spite of its low public profile, this tribunal carried tremendous importance for the well-being of Canadian telecommunications. Its supervision affected millions of dollars, and reached into a rapidly growing number of homes and businesses. What drove this culture, and what were its assumptions?

This culture entrenched strong ideas about the role of the state in telecommunications development and a regulatory practice consistent with those ideas. Ensuring a fair return on investment for Bell and determining equitable rates which balanced the interests of consumers, companies, and investors were the twin pillars of telephone regulation in this period. The defence of the public interest against a monopolistic public utility legitimized the process. Of course, the public interest by definition included liberal protection of private interests. On a practical level, this meant finding a fair way of dividing cost and benefit between producer and consumer. These principles formed the axes around which debate in the regulatory arena turned.

These regulatory principles were institutionalized in a series of legal and practical arrangements over the first three decades of the century. By the 1930s, the process was shaped by a few high-profile rate cases. What emerged by the interwar period was a type of split-the-difference adversarial quasi-judicial system that mediated between the demands of both producer and consumer. Bell would project its needs, determine the necessity for an increase in rates or a new stock issue, and apply to the board for relief. The board, in turn, would establish the need for a public hearing, and notify interested intervenors. The intervenors, mainly municipalities in the early period, would then assess Bell's application and prepare to tackle company plans for rate increases. Typically, the board would render its decision only after the hearings were completed, or else confirm a decision for interim relief. The board's principal mandate was to ensure that rates were non-discriminatory, just, and reasonable, but its reach extended beyond matters brought up in public hearings. Much of the regulator's work was determined by the minutiae of supervising the rate structures of the company, or approving a surfeit of tariff orders, exchange regroupings, and interconnection agreements with independents.

The board's legal framework may have been the cornerstone of the process, but this only partially accounts for the regulatory dynamic. Commissioners rarely used the full power given to them in apportioning the burdens and benefits of the public utility. The agency exercised its power with a light touch, creating a regulatory regime that steered clear of interference with Bell's managerial decisions. The commissioners were intent on the regulation of particular companies, for example, and not utilities in the abstract: a significant difference. As a result, they showed themselves sympathetic to the health of the company's financial structure, but less so to the grievances of those whose dollars supported that structure.

The board also used the concept of a "permitted level of earnings" to gauge the company's financial health. This was defined as what the company could earn from a rate structure, which would be the measure of whether rates were both just and reasonable. The level of earnings was a good benchmark of Bell's performance, and of its ability to raise capital in the marketplace. Successive decisions by the board demonstrated the astonishingly wide degree of latitude commissioners were allowed in arriving at a reasonable rate structure. This meant providing high enough rates to cover expenses and share dividends, and generate enough surplus to maintain the credit of the company in meeting new demands for service.

Arguments relating to social or economic conditions did not find a sympathetic hearing with the board. Ontario cities seeking lower telephone rates because of unemployment, for example, would get no satisfaction. Unemployment was not a sufficient condition to overturn the board's statutory obligation to oversee Bell operations. That meant limiting its scope to ensure no discrimination between subscribers took place, and little else.[6]

When the board was moved to act on a rate matter, stock issue, or complaints with the quality of service, the commissioners usually stressed regulative and corrective solutions over more intrusive managerial ones. Canadian regulators had considerable discretion in proposing those solutions, however, manœuvring with considerable freedom over the spectrum of regulatory issues. But that discretionary power sometimes came with unfortunate consequences, and could often result in a proliferation of public hearings, particularly in inflationary times.

The Canadian regulatory process also featured a political parachute for those really determined to escape judgment, since decisions of the board could be appealed to the federal cabinet under the provisions of the Railway Act. In order to forestall potential abuse of that avenue of appeal, the cabinet issued several orders-in-council between 1918 and 1933, under which board decisions could only be appealed if they had been subject to error. The warnings were successful in sufficiently narrowing that avenue. At the same time, the cabinet preserved its ultimate power over the regulatory process in reserving for itself the ability to vary, rescind, or alter any judgment of the board which could be defined as subject to "error." Depending upon the situation, the government could act to turn back a politically unpopular ruling, or one which did not take sufficient account of social and economic factors outside the purview of the regulatory agency. Once used, a different and more expressly political process was engaged. The potentially unpalatable political consequences made intervention an instrument used only in extraordinary circumstances.[7]

The BRC's 1927 hearings over Bell Telephone's application for rate increases was the last of its kind before the Second World War and the decision established the basis for much of Bell's rate structure. The board divided the company's telephone exchanges into nine groups and established the tolls for each group. The commissioners stressed that the business was one unit which operated in different places and had to be treated in that manner. Cost separation studies, which were the practice in the United States, were rejected as inappropriate to the Canadian situation. The precedents set in 1927 – that exchanges could be grouped for rate-making purposes – would be a prominent feature of telephone regulation in the postwar period.[8]

The heart of the exchange grouping system was not how much it cost Bell, but the principle of value of service. The board organized the company's rate structure around the number of telephones in a particular group: crucially, the individual exchange costs of providing the service were not considered. Put simply, the more telephones in a particular group, the more relative value the service would have for the individual subscriber. The value-of-service concept sought to eliminate the inequities that existed between services. Subscribers in Ottawa and Lachine, for example, both paid $2.57 per month for residential service, despite the Ottawa subscriber's connection to ten times as many telephones as the Lachine subscriber. In 1951 the board streamlined the process, making most reclassification automatic.[9]

The regulatory process for telecommunications was thus conditioned by a number of factors. Statutes established boundaries; decisions of the commissioners, their practical experience with rate cases, and board practice, entrenched style and method. Those factors were in turn conditioned by the wide, but not unlimited, discretionary power allowed the regulatory body. The result was a distinctive form of regulation.

The players reflected the nature of the arena itself. Lawyers, institutional economists, public utilities experts, bureaucrats, managers, engineers, and corporate officials flourished in the quasi-judicial environment of telecommunications regulation. But crucial to the process, of course, were three principal groups of participants: commissioners, intervenors, and applicants. Those who sat on the board were part of an exclusive legal, professional, or bureaucratic culture whose status was legitimized by the effectiveness of the regulatory process. The applicants by contrast were part of a corporate world who shared with the commissioners many of the same qualifications, but approached regulatory matters in a necessarily different way – as contestants. Finally, the intervenors, most often municipalities, who offered a contrast to the first two, met the quasi-judicial quality and nature of the regulatory process.

The appointment of commissioners to the board was primarily based upon a narrowly defined notion of competence. Some had served in Parliament or provincial legislatures. For all its politically affiliated membership, the board was remarkably free from the worst aspects of political patronage, irrespective of which party was in power. Commissioners such as F.M. MacPherson and Armand Sylvestre may have been good Liberals appointed by Liberal administrations, but partisanship had to be linked to some plausible technical capability in governing an adversarial process, while ensuring that judgments were consonant with public interest and the financial health of the regulated firm.[10]

The importance of the BTC for the regulation of transportation and communications was taken seriously enough to risk the grievance of backbenchers who believed they had a place on the board. When the Conservatives won power in the late 1950s C.B. Devlin wrote to Minister of Transport George Hees to express his astonishment that he had been passed over for appointment in favour of an insurance agent from Calgary. Devlin was a lifelong Conservative and clearly expected to be appointed. Though no stranger to patronage, Diefenbaker opted for Hees's recommendation of A.S. Kirk, director of traffic for the board and "possibly one of the two or three outstanding railway rate experts in Canada if not THE outstanding [one]." Similar incidents show that serving on the BTC required something decidedly more than a party membership.[11]

The cabinet also tried to ensure that commissioners were discharging their duties with probity and competence, as one commissioner discovered to his chagrin. W.J. Patterson had attracted the unfavourable attention of Minister of Transport Lionel Chevrier. The minister reported to St Laurent in August 1950 that he had found Patterson "unconscious on the floor beside his desk" from drunkenness. Patterson had also been making "an object of himself" in Union Station requiring the intervention of the station master. He resigned soon after, and was made lieutenant-governor of Saskatchewan, where his unconsciousness would pass unnoticed.[12]

These small incidents show that competence and capability were the key to board appointments when combined with partisan political interest. Both considerations were appropriately present in some combination, but governments were generally respectful of the high professional requirements that legitimized the performance of the board. The practice partially insulated governments from pressures to nominate local favourites, and reinforced the legal character and quality of the regulatory culture.

In its role as applicant, the Bell Telephone Company held the central position in the process. Given the corrective nature of federal telecommunications regulation, the company had a wide latitude in configuring its tariff structure and corporate organization and consistently sought to restrict the scope of the regulator's power. Bell also sought to minimize its contact with the board, especially in the setting of the public hearing. When called upon to do so, the company tenaciously sought to detail its reasons for increases, and defend its position against all comers. The company utilized not only its considerable legal resources, but also publicity and information campaigns to elicit public support for, or at least acquiescence in, rate revisions.

Opposition to Bell's plans fell to the cities and towns that lay within Bell territory in Ontario and Quebec, which became the voice of consumer discontent. Cities granted utilities local franchises;

telephone poles were erected and wires strung along city property as well. Rates were constructed entirely around the boundaries of city and town. Municipalities were also most susceptible to popular pressure from ordinary subscribers who wanted telephone rates to have the least possible impact on their pocketbook. The long and difficult relationship between utilities and municipal corporations in Ontario and Quebec was liable to become especially hostile with companies unmindful of local prerogative. Rural districts dissatisfied with the high cost and doubtful quality of rural telephone service were natural spokesmen for consumer concerns and ideally suited to carry the fight to the telephone company.

Federal politicians were infrequent but often significant players in the regulatory process. The quasi-judicial nature of regulation removed them from the main theatre of operation. Now and then, however, decisions of the board would become a political matter, especially if the judgment was particularly unpalatable. Once on political terrain, different considerations applied, and cabinet itself would decide the merits of the case, then return matters to the regulatory arena. By its very nature, political intrusion into the regulatory mechanism would be temporary and usually not offer any long-term equilibrium solutions to the problems encountered in the process.

The defence of the interests of both applicants and intervenors were of necessity filtered through a phalanx of lawyers well versed in administrative law. The fact that the largest telephone utility in Canada was based in Montreal likely generated a spin-off effect in terms of legal specialization. Naturally, Bell Telephone retained some of the finest practitioners of the craft throughout the period – Norman A. Munnoch, J.L. O'Brien, F.P. Brais, F.A. Burgess, and A.J. de Grandpré to name a few.

Even when the municipalities desired competent and effective legal representation before the board, their first stop was Montreal. Although municipalities often sent their own solicitors to the hearings, a Montreal lawyer, Lovell Caverhill Carroll, led the opposition to the company. Carroll was solicitor for the town of Mont-Royal, and professor of commercial law at Sir George Williams College. This mild-mannered lawyer, who operated without any assistants or secretarial staff, would become a familiar feature of Bell rate cases throughout the 1950s and 1960s.[13]

Lawyers were the performers; through them, the regulatory process would derive its tone, and filter much of its substance. Since the process left much to the discretion of the commissioners, the probity and proficiency of counsel to cajole, plead, persuade, and enlist the evidence to their clients' cause was a predominant feature in the regulatory dynamic.

Taken as a whole, regulation was demonstrably the intersection point between private and public power. By helping to define the boundaries between the two, regulation shaped the agenda of public discourse. The culture steered public debate over telecommunications into an environment where applications could be measured and evaluated by experts against judicial conceptions of equity, respect for property rights and statutory responsibilities of the state, and insulated from popular pressure for lower rates. The public interest was precisely defined to include not only minimum rates, but also financial security for Bell. The result often, but not always, favoured the interests of Bell. The board also mediated between consumer and producer in good faith, and legitimized the existence of monopoly control. This was the regulatory culture that extended into the two decades after the Second World War. Once there, regulation would encounter a telecommunications structure on the threshold of momentous institutional expansion.

A CONTESTED DEVELOPMENT

Bell's plans for expansion and development reflected Canada's remarkable postwar growth. These plans, however, aroused considerable opposition. Throughout the 1950s, the utility's traditional adversaries – the municipalities of Ontario and Quebec – would confront Bell management in the regulatory arena. Both growth and dissent would affect the balance of power between agency and utility.

After 1945, wartime savings and investments were released into the hands of consumers and provided a firm foundation for an outburst of economic activity. The result was a remarkably large demand for telephone service. By the time the company had appeared before the board in September 1946 for a new stock issue, the backlog of construction and telephone demand had reached impressive proportions. Bell directors used the postwar vocabulary of "reconstruction" to describe their plans for telecommunications expansion. This was accompanied by a vigorous publicity campaign to maintain the momentum of demand (see illustrations 1 and 2). The company managed to raise $19.2 million for 1946 and over $54 million in 1947 from two separate stock issues. The only opposition to Bell's request had come from one postcard sent to the chief commissioner asking the board to deny the stock issue until the company "pay the millions of dollars plus compound interest that was fraudulently withheld from the shareholders."[14]

The company must have dreamed of that postcard when the company applied for a general increase in rates in October 1949. The last time Bell had asked the board for a raise was in 1926. The basis for the application was the demand for telephones. Bell had argued that new telephone services did not necessarily result in increased revenues. Pent-up wartime demand had resulted in overloaded plant, and a public no longer content to be patient in its demands for new or upgraded service. Its postwar construction program added half a million new telephones, but supplies still fell short. Bell proposed to spend $141 million in new construction between 1950 and 1952 alone. "Courageous planning" was called for. Tables 4 and 5 illustrate the impressive growth in demand for telephone service and its supply between the prewar and postwar periods. In every area, both demand for the hardware and its use exploded in less than a decade.[15]

If Bell management was expecting the board to expedite matters, it was soon disabused of the notion. Bell had applied for immediate relief that would give it $9.3 million of the $23.8 million asked for in the main application. Bell counsel Norman Munnoch reminded Montreal city council that in that city alone 20,228 orders for new service remained unfulfilled, while demands for upgrade exceeded 26,000. Bell lost on both counts. Chief Commissioner Archibald ruled that the hearing would be postponed until March 1950, and the immediate relief would also have to wait until the respondents could have a chance to file answers to Bell's main application.[16]

Bell's application met with organized resistance from its traditional adversaries, the larger municipalities of Ontario and Quebec. The larger cities were particularly impatient with the application, since they had experienced the most popular pressure to hold the line on telephone rate increases. The average subscriber paid about as much for his telephone as for electricity. As a telephone service cost a couple of dollars per month, the consumer looked to intervention by the board to keep that cost down. Communications were essential, so the costs were naturally linked to the rising cost of living, especially in Montreal and Toronto. The application would mean a 20 per cent increase in rates, with customers in the two cities absorbing most of the increase. Generally, prices in urban centres had remained fairly stable until the end of 1946. The two following years saw significant cost-of-living increases of 9.6 per cent and 14.4 per cent. By 1950, prices were more than two-thirds higher than they had been in 1939.

On 8 March Lovell Carroll led the municipal attack on Bell by asking for an immediate dismissal of the application, on the grounds that the company had failed to reclassify its exchanges into the proper rate groups, and because Bell could put its $12.5 million surplus to work on construction instead of giving it away as share profit. Carroll contended that the application was merely one of a pattern of applications Bell made after major wars "at a time where there is a terrific ... and abnormal and unnecessary demand for telephones and when wages and cost of materials ... are high." Although the board refused to dismiss the proceedings, Carroll's opening salvo put Bell on notice that the municipalities were prepared to oppose the application with great vigour.

The opposition plan was based on a multi-pronged assault. Criticism fastened upon Bell's extravagance in everything from its construction program to its non-contributory pension plan for its employees. Carroll took every opportunity to deride Bell spending and provision of service as "gold plated." The company's ambitious construction program in recent years offered the respondents lots of ammunition. Gross construction had only risen slightly during the war years, from $14.8 million in 1939 to $17.9 million in 1945. The real story, however, lay in the postwar numbers: in 1946 the company spent $33.7 million, reaching a sustained plateau in 1948 of over $82 million annually. The results bore testimony to the expense. In 1939 the company had three-quarters of a million telephones; by 1952 it expected to have almost two million. Sustained demand and the consequent need for construction and modernization would become the centrepiece of Bell's regulatory strategy. Table 6 shows the impressive expansion of Bell's construction program.[17]

Another area that concerned the municipalities was the ratio of debt to equity, and it would become a familiar part of the municipal critique of Bell in subsequent rate cases. The municipalities argued that a 60 per cent debt ratio could be safely managed. The issue was especially important to the cities since the financing of the construction program might be borne more from increasing the debt ratio and less from general increases in rates. In any event, they claimed, the company had overstated its difficulties and had become addicted to spending. Even the condition of the restrooms at Bell offices was enlisted to prove municipal charges of Bell's financial profligacy. In any event, the "amazing lack of cost records" prevented a proper rate setting.[18]

Advertising spending had increased over 50 per cent since the wartime period, prompting Carroll to ask Bell's advertising manager if the company was not insulting the subscribers' intelligence to tell them "how to pick up and put down the receiver." The company might be more financially sound if it devoted its advertising expenditures to some other purpose. Current maintenance costs were shown to be neither justified nor economical. Carroll offered the same assessment of the salaries of non-union personnel within the company, a position echoed by other counsel on the issue of pension expenses.[19]

The board took over a month to decide whether to grant the interim increases asked for. Bell President Frederick Johnson was called to describe the ill effects of not granting the relief on company credit and telephone service. The commissioners responded in July 1950, and approved almost all of the increases the company had asked for. The judgment produced an extra $1.5 million in extra revenue, $.5 million short of what the company asked for in October 1949.[20]

Bell had received most of what it asked for even before the final judgment had been rendered. But in the process, the company had had to bear some intense scrutiny and even occasional bruising by the municipalities. Carroll reminded a resentful Bell that if there was a choice between a possible injury to the company and a definite injury to the subscribers, "it was a matter of public order that the public good must be served."

Bell counsel pounced on Carroll's pronouncement and for good reason: if there was any hint that the cities were speaking for the public, Bell's position could be seriously compromised: "Now, not only do I say that the municipalities here have no right – or authority – to speak for the public as they propose to do by any suggestion of that kind that ignores that branch, or part of the public that have not got telephones – those are not being spoken for by any of our friends."[21]

The final decision on Bell's application was rendered by Commissioner Wardrope on 15 November 1950. The board granted virtually all the increases the company had asked for. It argued that its decision had to be reasonable in providing adequate protection for the company, and especially in safeguarding its ability to attract capital for the tremendous demand for service. The municipalities had defined the parallel interests of subscribers and management in a much narrower way. The "reasonable zone" had been clearly defined to exclude the municipalities' conception of what was reasonable. Both depreciation, pension and current maintenance costs were found to be reasonable by the board, as was the relationship between Bell and AT&T, and between Bell and Northern Electric. The municipalities had argued that the services rendered by AT&T were not worth what Bell was paying for them and were not based on actual and ascertainable costs.[22]

Reaction to the increases was not surprising. The municipalities had fought through most of 1950 to convince the board to hold the line on the increases, and the regulator had failed to deliver. In Toronto, Mayor Nathan Phillips instructed the city solicitor to canvass opinion among the opposing municipalities about future strategy. This call was particularly apt, since circumstances would force Bell back to the board less than a year after it had given it an increase in rates.[23]

The company's 1951 application asked for another 10 per cent. The company had suffered from the inflationary jolts in wages, taxes, and host of other costs that had accompanied the Korean War. In all, the increases would amount to $15.8 million in additional revenue, most of which would come from another general increase in rates. The cost of a telephone would jump from $12.00 to $15.00 per month for a business telephone, and $4.75 to $5.50 for a residential line. Long distance rates would not be affected, reflecting Bell's preference for the certainty of local service revenue over the risk of the more demand-elastic long distance market. The company also suggested that the normal rate of dividend must be maintained to yield earnings of 43 cents per share; the stock performance of the telecommunications sector had been stable, and Bell wanted it kept that way. The two-dollar surplus per share had been reduced to 26 cents. Wage settlements had increased by $16.7 million, or 27 per cent above the 1949 figure. Circumstances required immediate interim relief.[24]

The 1951 rate hearing was a rerun distinguished only by the fury of municipal opposition. The Canadian Federation of Mayors and Municipalities questioned the company's interpretation of what was "reasonable" and "normal." The federation decided to target its opposition: rates to cover the Defence Surtax imposed on companies as a result of the Korean War would be overlooked. Other parts of the application, however, were tantamount to "guaranteed protection" for the company's financial operations. Over Bell objections, the federation urged another full discussion of the necessity of the construction program, and the possibility of increasing the company's debt ratio, so that financing could be at least partially met by methods other than raising rates. Bell was not going to get away with two increases in two years so easily.

In the meantime, in September 1951 both the Confédération des Travailleurs Catholiques du Canada (CTCC) and the Trades and Labour Congress called for the nationalization of the telephone company. Le Travail, the CTCC weekly, accused the company of using the rate increases to increase its dividends to shareholders, not improving service for ordinary subscribers. The calls for public ownership were quickly rebuffed by the mainstream press across the country.[25]

The board's 1951 decision permitted an average debt-equity ratio of 40 per cent as well as earnings of 43 cents per share surplus. "We are, under present circumstances," Commissioner Wardrope wrote, "somewhat impressed with the desirability of strengthening a utility's financial structure to the extent that under conditions of a less favourable money market it would not be basically handicapped in securing necessary additional capital at reasonable terms."[26]

The first two Bell rate cases of the postwar period contained a mix of the traditional and the modern. If the faces were different, many of the issues were the same: depreciation, long-term financial security for the company, and municipal complaints about Bell's extravagance and gold-plated visions. What was new was the sheer magnitude of the demand for telephone service, with no end in sight. The question of the balance of power between agency and utility seemed irrelevant in light of such massive demand: people wanted it and were willing to pay for it. In this context, the case of the municipalities seemed almost ungracious. Bell had to be given a free hand to extend and diffuse the telecommunications network. Unprecedented demand was also having significant effects on the balance between Bell and the BTC. Put bluntly, the board was beginning to find it hard to keep up with the pace of change.

BUILDING THE TELECOM NETWORK IN THE 1950S

How did Bell and the BTC manage the rapid expansion of telecoms in the 1950s? Through such mechanisms as extended area service and the regrouping of telephone exchanges into higher rate groups based on the number of telephones, the company was able to pay for its high rate of growth and diffuse the telecommunications network quickly and efficiently. The BTC offered its support to these plans, but, often over the objections of municipalities, whose constituents bore the cost of expansion.

Throughout the 1950s, the board managed to regulate Bell outside the glare of public hearings. Both Bell's exchange groupings and its network expansion plans were municipally based. Metropolitan or extended area service was just one such phenomenon. By extending its free calling areas in metropolitan districts to include suburban centres, Bell could offer more subscribers more free calls while charging higher rates. The company subsequently polled its subscribers in Ottawa, Toronto, Quebec, Windsor, and Hamilton to determine whether subscribers were willing to pay the higher rates associated with larger exchange groups. Each major area in turn consented, with huge majorities in favour of the extended service plan. In Aylmer and Gatineau, across the river from Ottawa, for example, over 95 per cent of subscribers voted, and 90.6 per cent approved of the plan. The board acknowledged that subscribers in the urban area itself would either be disinterested or opposed, since the Ottawa exchange would outgrow its limit sooner. On that basis, the board withheld approval for the plan in order to protect the subscribers in the Ottawa rate base area. Similar plebiscites were held in the Quebec City and Windsor areas, with largely similar results.[27]

Exchange groups were another matter: every subscriber in Bell territory would be affected sooner or later. Cities and towns were continually transferred to higher-cost exchanges, reflecting the comparatively rapid rates of urbanization in Ontario and Quebec. Growth did not stop objections over regrouping exchanges from flowing to the board, however. The character of the objections were familiar: city councils and subscribers viewed "extensions" such as private branch exchanges and similar installations as distorting the true value of telephone service. As such, the groupings were "purely arbitrary and unfair" to many municipalities with a high proportion of extension telephones in the total count.[28]

Secondly, urban subscribers were continually irritated that they were "required to pay increased rates because of the Suburban development." Third, Bell often applied for regrouping in an arbitrary manner. For example, Bell applied for changes affecting the Quebec City area that resulted in a 24 per cent increase in residential service and a 69 per cent increase in business service – all because the company had waited until the region jumped two categories. The board agreed on the facts of the case, but declined to cushion the blow all the same. Municipalities had persistently argued before the board that the way Bell handled exchanges led to large exchanges subsidizing the smaller ones, without any reference to the individual costs of providing the service. Other municipalities objected to the arbitrariness of the sizes of the rate groups. Trois Rivières objected that the transfer to Group 7 would force users to pay "as high a rate as if there were more than double the number of telephones in use." Time after time the board reaffirmed the principle of the distribution of the rate burden according to relative value of service. A score of municipalities who found themselves in a higher grouping, but pleaded hardship or unfortunate economic conditions, were rebuffed. In judgment after judgment the board ensured that no discrimination between areas was allowed to occur in strict observance of its statutory mandate.[29]

In the larger cities, however, the Bell referendums showed that suburbanites ardently desired rapid connection. The board accepted the plebiscites with satisfaction. In the Montreal case, only twenty-four out of almost half a million subscribers in Group A opposed the plan. Increased rates and charges would naturally result from the extended area service plan, as was the case with Toronto in 1955. This mechanism for legitimizing the company's plans for extended service had the virtue of simplicity: subscribers would vote, return large majorities in favour of the service, and the service would be established. The plebiscites were an ingenious way to expand the service, but by design did not include urban subscribers who were decidedly more cool to the idea of adding tens of thousands of exchanges to their existing group. This was especially true since Bell could apply for exchange group increases at will.[30]

The board also received numerous complaints arising out of the company's regrouping because of the inadequacy of the service in an area. Protest was usually situated outside urban areas, where service upgrading had not been a priority for the company. Residential subscribers in one Quebec municipality complained that the absence of single party lines was greatly frustrating communications: "Il lui faut ainsi attendre parfois une demi-heure, parfois trois-quarts d'heure, et même parfois toute la journée." Quebec subscribers and municipalities voiced more protest than their Ontario counterparts, partly because system modernization lagged behind. As tables 7, 8, and 9 show, telephone expenditure in Quebec was less than that of Ontario in the 1940s; however, the reverse was true by 1959. But by that time, the Bell network was more diffuse in Ontario than in Quebec. This could have been a double frustration, since Québécois were paying more for telephone service, and getting less value for it.[31]

In spite of its problems, exchange regrouping and the extended area service strategy served the company well. It did so by entrenching some long distance in local service revenues, which were the more stable, while making suburbanites happy. The machinery set up or suggested by the BTC in the form of orders respecting exchange regrouping or plebiscites had allowed the company to grow in tandem with demographic surges in both urban and rural areas. Subscriber plebiscites gave technological expansion the aura of popular suburban sovereignty, while at the same time minimizing the outlets for urban dissent. Bell polling continually reported enthusiastic support for the planned conversion of most suburban areas to extended service. Perhaps there was not much dissent in cities. But this is immeasurable since city dwellers were not polled. Bell succeeded in not only providing service, but also integrating its system more fully into the urban-suburban landscape. The arrangement also provided Bell with a steadily increasing source of revenue which accentuated its concentration on more lucrative markets in its territory.

Demand for telephone service and upgrading nourished Bell's expansion. The company found itself with high demand that promised healthy returns and ensured continued construction. Larger urban centres were mushrooming. Trends for the telephone industry were doubly encouraging; a much larger proportion of consumer expenditures was devoted to electricity, gas, and telephone services than ever before. The supply of telephone service could not keep up with demand. In 1941, for example, one million homes had telephone service; in 1955 the number had almost trebled, to 2.7 million. Indeed, expenditures on telephone service had risen approximately fivefold between the late 1920s and the mid-1950s.[32]

Appearing before the Royal Commission on Canada's Economic Prospects in 1956, Bell President Tom Eadie underlined the telephone's contribution to the national economy by noting that more telephones had been added in the past ten years than in the previous seventy. At the same time, Eadie could assure the commission that improvements in transmission and switching were resulting in major improvements to the telecommunications network. The microwave radio-relay network under construction would meet expanding national requirements as well. The phenomenal growth of telephony told the story best. Of course, the maintenance of the golden age was based on the continued "reasonableness" of the regulatory regime.[33]

The telephone industry in general, and Bell in particular, enjoyed a special prestige in the iconography of progress. As J.K. Galbraith remarked in 1956, resource development, geological investigation, and transportation and communications were exalted economic activities, especially in Canada. While time and economic growth would make returns from these sectors less dramatic, they continued to enjoy an elevated importance. Telecommunications had a public relations edge over the other sectors, since its technological cachet linked Bell to the vanguard of economic activity. As illustrations 1 to 5 show, Bell did not hesitate to take advantage of its position as providers of technological and economic progress.[34]

REGULATION UNDER ATTACK:
POPULISM AND TELEPHONE RATES

During the greater part of the 1950s high demand and robust financial health had put Bell in an excellent position. However, by 1957 economic slowdown and an increasingly restive public irritated at high costs combined to create serious problems. The happy conditions that prevailed for most of the 1950s could no longer be taken for granted.

Consumer prices began rising in mid-1956, and economists predicted a slackening in the rate of economic expansion. Yet Bell seemed insulated from the downturn. In order to meet the massive demand, the utility pushed for more capitalization by revision of its Special Act. In August 1957, Bell also applied for increased tariffs. The request would increase the company's total operating revenues by $24.2 million per year. Demand had risen in almost every category of telephone service provision, but expenditures had outstripped growth. Both local service and long distance revenues had doubled in six years. Innovations in long-distance switching technology improved speed and accuracy, but savings were partially offset by maintenance charges, depreciation, and capital expenses. The application asked for a permitted level of earnings of 7.7 per cent on the equity per share, a figure that would amount to $2.65 per share. On the matter of debt ratio, the company held firm to the traditional threshold of 40 per cent that management considered necessary to meet its capital requirements and demand.[35]

Bell's application again encountered stiff opposition from the municipalities. Lloyd Jackson, the mayor of Hamilton and president of the Canadian Federation of Mayors and Municipalities, viewed the application as a declaration of war. Jackson envisaged an "all-out effort" to fight Bell that would include the best legal counsel and expert testimony. The larger cities decided their case would have more impact if all the municipalities over 10,000 people in Bell territory would oppose the increases. The case would not be easy to prove: Bell's growth had been substantial.[36]

By the first day of hearings, many municipalities in Ontario and Quebec had joined the larger cities in opposing the increases. Lovell Carroll appeared on behalf of thirty-six municipalities, and predictably charged that the company had failed to demonstrate its need for an increase. Carroll was satisfied that the company had made its case only in the area of wage and tax increases. Otherwise, Bell was making a claim for future expenses "which had not yet materialized." As a result, the burden should not be imposed on telephone subscribers. If the company maintained a prudent capital structure, then a 45 to 50 per cent debt ratio combined with a $2.27 earnings figure per share would be more than sufficient for its needs. Bell's "naked claim for higher earnings" had to be stopped. "If there is any threat to the company's ability to meet the $2 dividend payment during the next few years," Carroll concluded, "it is a threat caused by its own imprudent financial policies – not by the present telephone rates." The United Electrical, Radio and Machine Workers presented twelve thousand protest cards opposing Bell's application, and especially urged the inclusion of Northern Electric in the consideration of the rate base. Those arguments were also echoed by Eugene Forsey, representing the Canadian Congress of Labour.[37]

The large number of written submissions opposed to Bell's application underlined labour's concerns. Resentment of Bell lingered over hidden profits, non-telephone services subsidized by the subscribers, the poor quality of service provided, and extravagant management. For many, the company had already increased rates in their particular region with upgrading of exchanges. In all, the opposition case was summed up by George Mooney's hopeful report to Toronto City Council: "While it would be imprudent to prejudge the outcome of the case, I think we can justly claim that a very strong argument against the rate increase is being presented."[38]

The judgment delivered by Chief Commissioner Clarence Shepard dashed any hopes of municipal victory. The board approved a host of rate changes to both local and long distance services. The judgment satisfied no one: Bell obtained 43 per cent of its demands, or an increase of $10.3 million. The company also managed to obtain a 7 per cent return on total capital which would yield $24 million. That translated into a 3 per cent increase across the board. The commissioners also reaffirmed the company's traditional 40 per cent debt ratio. Even so, Thomas Eadie, the most energetic proponent of expansion, viewed the decision as a threat to that growth, and hence to Bell's financial position.[39]

The decision that was to cause the most controversy for the board was its ruling on Bell's deferred credit – income tax account. This was no esoteric matter, since Bell had accrued over $12.5 million in deferred credit during 1957 alone, for a total of $47.9 million that was not part of the rate base. Changes in the Income Tax Act in 1953 had made it possible to charge depreciation on the basis of capital cost allowances authorized under the regulations, regardless of the amount of depreciation which may be recorded in the company's books in the same period. The tax paid was thus often less than charged on the company's books. This was known as the diminishing balance method.[40]

Bell managed to convince the board that not to apply this method constituted an "improvident act of management." The company suggested that if the board accepted the arguments of the municipalities on deferred credit, it would impose on Bell the risk of prophesying, while leaving the company in a position where it would be unable to protect itself against the risks to development if it guessed wrong.

The value of a deferred credit account could be substantial. It all depended on Bell's alternative use for the funds. The company could plough the savings into high return assets, or speed up its construction program.[41] This was not a novel practice for a large Canadian firm, but Bell was no ordinary Canadian company. In this situation, its status as a public utility had turned what was legal and economically efficient into something that was allowed to stand as the symbol of its monopolistic arrogance and desire to profit excessively from its subscribers. The board's approval broke new ground, since no established pattern had been set either in Canada or the United States about the question of deferred taxes for rate-making purposes. Provisions for deferred taxes were thus allowed in calculating net income.

The issue of the deferred credit – income tax account was at the core of municipal objections. The account had swelled to $47.9 million between 1953 and 1957, and was expected to rise to over $61 million in 1958. The company was forcing "customers to make an involuntary contribution to capital." The result was a permanent tax savings that would benefit the company, and leave the subscribers with no benefit at all.

A similar judgment on tax and depreciation by the board in a railway case in December 1957 prompted eight provinces to appeal to the governor-in-council. The appeal resulted in the deferment of any railway rate increases until the cabinet had considered the appeal. Since the principles were the same, the municipalities took the lead and appealed on the telephone case. In fact, the government had contemplated suspension even before the BTC had rendered a judgment. Chief Commissioner Shepard cornered the clerk of the Privy Council, R.B. Bryce, at a dinner party on 8 January 1958 to ask what the intentions of the government were with respect to the Bell case, given that the railway rate decision had been suspended on the same principles. Bryce thought about it, and soon after advised Prime Minister Diefenbaker that action would be necessary to suspend the decision of the board on the Bell case as well. "The essential principle in the railway rate case," he advised, "is the recognition of the need to place funds in a reserve against future income taxes as a proper charge in rate making." Letting Bell do it and not the railways would not be consistent. The suspension would have the added political bonus of the state immediately seizing the initiative, since an appeal was anticipated in any event.

The difference between the railway case and the telephone case lay elsewhere. Opposition to Bell could not hope to muster the kind of support and vehemence that was characteristic of provincial and business opposition to railway hikes. The structures of the transportation and communications utilities were quite different: rail transport featured few units, while communications were characterized by extension and diffusion to hundreds of thousands of consumption points. The politics of railway regulation had is own unique dynamics and set of actors. As Bryce noted shrewdly, "The main effect [of the telephone increases] is on the consumers and they are notoriously complacent in the defence of their interest."[42]

The telephone rates case could not have come at a worse time for Bell. The board's decision was announced on 10 January 1958, near the beginning of the new session of a Conservative minority parliament. Diefenbaker's government had moved quickly in the first months of its mandate, passing a number of popular measures from price support for turkeys to $150 million for loans on low-cost housing. Even before the board could announce its findings, the government had taken the decision to suspend the increases, at the very least.[43]

The Diefenbaker government was re-elected in a landslide victory barely months after his 1957 election to power and the appeal against the telephone increase would be heard two weeks later. Bell pushed hard, although the odds were against the company. Its memorandum to the government was a mixture of reference to legal precedent and heaping disdain on much of the evidence and expert testimony of the municipalities. The company was being asked to "risk the future financial stability of its enterprise" and leave present-day users worse off. Dark consequences were predicted.[44]

On 29 April the prime minister announced that the government had decided to rescind the rate increases. Order-in-council 1958–602 directed the board to no longer view credits to tax equalization reserves as necessary expenses in determining rates. Two reasons were given. The Diefenbaker cabinet first cited the uncertainty about whether the reserves would ever be used for tax payments. The decision also cited the unfairness of imposing the full cost of a "distant and uncertain contingency" on subscribers. Both Bell and the BTC were dealt a significant setback.[45]

The denial of the increases was the first action of its kind in thirty-five years. The motivation in annulling both telephone and freight rate increases was to halt a wage and price spiral that had been a chief concern in the campaign of 1958. The *Globe and Mail* reported that the announcement had come as no surprise although "the fact that a federal cabinet would take such an attitude was a matter for astonishment."[46] The prime minister had made no secret of his opposition to the rate increases, and was even willing to do something about it. This was a noticeable shift from the previous government's arms-length view of government-agency relations.

The decision served notice, the *Gazette* claimed, that "the Government formerly headed by a Quebec corporation lawyer, Mr St Laurent, is now commanded by a Saskatchewan jury lawyer "who shared western alienation toward big interests." Lloyd Jackson enthused that the municipalities had "won all the way down the line" with the rescinding of the board's decision.[47] The move was widely interpreted in press accounts as anti-inflationary and pro-consumer.

The reversal was a serious blow to the prestige and authority of the Board of Transport Commissioners. The reversal of the decision was a forcible reminder that the public interest had not been served. The elite consensus among corporate officials, commissioners, and experts over the proper regulation of the utility had become isolated from the public interest. The appeal engaged a political process which demonstrated the limits of the regulatory mechanism, and served as an early warning to the board that it could not escape political consequences of regulatory actions.

Bell President Tom Eadie must have believed that the company had not pushed hard enough. Shortly after the order-in-council, he spoke with the prime minister and explained the "difficult position" in which the company had found itself as a result of the recent decision. The maximum development of the telecommunications network was placed in direct jeopardy. Eadie attempted to clear the way for another application for rate increases. "Is it fair to ask", Eadie queried Diefenbaker, "that if we pay our income tax what the Government's attitude would be in the event that another application was made to the Transport Board for an increase in rates and the Board granted the application[?]" Diefenbaker promised to consider the matter and to give Eadie an answer.

The answer must have been a positive one: four weeks later, in June 1958, Bell applied for another round of increases. Later, and more directly, Eadie implored Diefenbaker not to interfere: "A repetition of that experience," he warned, "would be devastating ... [t]he resulting curtailment of work would deprive 10 to 12 thousand Canadians of employment ... Years would pass before Bell could recover from another adverse judgement. Bell could only hope that the Conservatives would get the message that another 'disagreement' must not be allowed to occur.[48]

Bell was now asking for a $16 million increase, citing the usual costs as well as the disappearance of the deferred credit. Several provincial governments decided to intervene, anxious to head off a precedent over the question of deferred taxes and possible linkages to rate-making. They joined the municipalities in arguing that Bell's rate structure violated the spirit of the order-in-council.[49]

Opposition to the increases involved a broad cross-section of interests, but generally speaking resistance had a populist base. Conservative Stanley Petrie seemed to sum up the general view of Bell and its second application in his letter to Diefenbaker: "Many people hold little brief for the Telephone Company. They are taped as being highhanded, dictatorial, [and] monopolistic … I, along with many of my friends, hope and trust that the answer will be no [to the increases]." Eugene Forsey was more blunt, charging that Bell just couldn't take "no" for an answer, launching a fresh attempt "to get what the government of Canada has just decided it should not have."[50]

The municipalities were quick to resume their fight with Bell. The critical matter was, again, the method of depreciation. This time, company strategy short-circuited municipal battle plans. Bell had deliberately decided to implement straight-line depreciation rather than the flow-through method. In one stroke, Bell doubled its tax pay-out, but would be able to recover the loss from subscribers. For rate-making purposes, the practice was objectionable, but legal. Carroll was reduced to the frail argument that order-in-council PC 1958–602 had, by implication, inference, and result, favoured the flow-through method and nullified the need for a general increase. The argument was a crucial one, not only from the municipal point of view but from the perspective of the legitimacy of regulation. An exasperated Carroll warned of the consequences to the board itself if Bell got its way: "Regulation would be impotent if it fixed rates on the basis of what Bell wanted to do rather than what it should do." The answer lay in increasing its pay-out from Northern Electric, reducing its contributions to the pension plan, and "further economiz[ing] like all other companies have been obliged to do in 1958."[51]

The company mounted an all-out public relations offensive to complement its Ottawa lobbying. There was nothing subtle about Eadie's testimony before the board. He gave extensive evidence of the significant contribution that Bell had made to the Canadian economy, national defence, and in the life of ordinary people. Bell had embarked upon a construction program worth well over $1 billion in the postwar period; from 1955 to 1958 the outlays amounted to $625 million, an amount equivalent to the Canadian share of the St Lawrence Seaway project. The company had not asked for a rate increase since March 1952, yet both demand and costs were rising. This time, Toronto's Board of Trade supported Bell's application, fearing deficiencies in telephone service which would result from a denial of increases. Bell had not hesitated to mobilize its employees to the cause. They were instructed to sell the merit of the rate increases, and to pressure city councils who had opposed the first round of increases. In Ottawa, an influential advisor to the prime minister, Senator W.R. Brunt, wrote Diefenbaker to discuss the renewed Bell application.[52]

Unlike a few months before, significant editorial support accompanied the renewed application. In an editorial appearing shortly after the application was filed, the Toronto *Globe and Mail* scolded those opposed to Bell for risking "throwing thousands of jobs down the drain in order to save the voters one cent and a half a day on their telephone bills." The "wage-tax-price" spiral was regrettable, but Bell's intentions were honest and honourable. The *Financial Times* chided meddlesome provincial and municipal leaders for being more concerned with curtailing legitimate expenses of business, and less with "curbing the financial demands of the state."[53]

The greatest worry reflected in press accounts was that Bell would not keep pace with technological development because of political tampering with the board's decisions. The company also might not be able to sustain its big construction program "designed to keep step with the growth of Ontario and Quebec communities." The government's objective in holding inflation down was laudable, but should in no way interfere with expansion. Editorial and press comment reflected the regulatory culture's success in promoting concern with technological diffusion and development. Many were willing to pay higher rates if it meant preserving the integrity of Bell's construction program, no matter how gold-plated it was. Failing to do so would put the system in jeopardy.

The flood of editorial comment on the case failed to move Toronto city council. Under the leadership of Mayor Nathan Phillips, in the summer of 1958 the city moved to oppose the second set of rate increases. Aldermen charged that they had been subjected to a "high pressure public relations campaign by the company." The mayor claimed that Bell had incited its shareholders to bring pressure to bear on council and its officials had visited city aldermen to persuade them that the increase was justified. The press, hostile to council's opposition, called it a declaration of war on the telephone company. The *Financial Times* saw nothing wrong with company officials and shareholders lobbying council: "the company could hardly be expected to do other than use every means in its power to smooth the way for its renewed application. Shareholders – about 60 per cent of whom, incidentally, are women – would hardly take much stirring up."[54]

By mid-August, forty-four Ontario and Quebec municipalities announced their intention to oppose Bell's application. George Mooney, executive director of the Canadian Federation of Mayors and Municipalities, was not unmindful of the criticism the municipalities had incurred in the course of their opposition, and defensively told the press that "municipal governments are not opposing the Bell application on specious grounds. They are doing so because they believe it is as much in the interests of the company as well as the subscribers to subject the application to full and careful scrutiny."

The editorial staff of *Le Droit* supported the action on the basis that no explanation of Bell's increasing profits and declining service had been offered. "We do not dare [say] that the company be nationalized, but we refuse to approve the increase without understanding the need for it."[55] It was in this environment that hearings began on 16 September 1958 before Chief Commissioner Shepard, F.M. MacPherson, and H.B. Chase. The intervenors quickly got down to business in debating Bell's questionable handling of depreciation and income tax matters.

The board issued a judgment shortly after the hearings, approving in full the company's application for $17.5 million in additional revenue for 1959. The commissioners ruled that it would not interfere with Bell's decision over how to pay its taxes. Tom Eadie publicly thanked the board both for the decision, and the promptness with which the judgment had been issued.[56] The rates were to go into effect in November 1958.

The case was immediately appealed to cabinet, as the previous decision had been. The cabinet would hear petitions not only from Bell, but also from municipalities, provinces, and labour groups. The company was quick to lobby the prime minister and launch an aggressive advertising campaign. Bell was banking that another rates rollback would create enemies for the Conservatives far beyond the narrow question of telephone rates.[57]

The situation had changed in the eight months since the cabinet had taken up the issue. Its springtime ardour gave way to an autumn ambivalence. Rescinding another rate increase could result in lay-offs; letting matters stand would bring the government $13.4 million more in taxes and would protect Bell's ability to raise capital. The company was of signal economic importance in a Canadian economy troubled by recession. Ultimately, the cabinet disallowed the appeal and upheld the decision of the board. The Diefenbaker government was convinced that the BTC decision struck some balance between the telephone user and the interests of the company, and it was anxious to avoid more criticism over its handling of economic matters.[58]

This experience left Diefenbaker considerably chastened. Intervention into the complexities of the regulatory mechanism held as many perils as potential political advantages. The Conservatives even contemplated removing cabinet's veto power in rate matters, at the suggestion of Tom Eadie. Diefenbaker was intrigued by the idea, and referred the matter to George Hees, the minister of transport. Hees was decidedly more cool to the idea. Although he admitted the creation of a cabinet-proof machine would protect the state from political pressures over rates, government had to retain some power – the issues at stake were often too important to be left to the regulatory authority. Given the problems confronting the Conservatives, Hees was content to let the matter drop.[59]

The reversal of the board's judgment and the flood of controversy that followed drew the participants in telecommunications regulation into unfamiliar and expressly political territory. In not recognizing the implications of respecting managerial autonomy over the issue of income taxes, the board exposed its remoteness from considerations beyond its administrative imperatives. The economic slowdown that began in 1957 sharpened the populist eye on public utility firms such as Bell, who had seemed to profit even in recessionary times. Put simply, it did not seem fair. In that context, the case presented by the cities within Bell territory had a particular resonance. Arguments against Bell's efficiency and extravagance were more effective in 1958 than they were in 1950. To their good fortune, populists found a champion in the Diefenbaker Conservatives. The 1958 débâcle threw into question the ability of the BTC to balance public interest and private dividends. The stress on network expansion had clouded the issues. In the process it served to shift the balance of power between agency and utility, and created serious health implications for the board.

The ultimate outcome one year later did not cause any lasting damage to the ethos of expansion and had little effect on Bell operations. Resistance to the company's grand construction program faltered, overshadowed by the technological and economic benefits telecommunications would bring. If politicians and subscribers were wary, Bell and its allies were prepared to defend, persuade, convince, and even threaten if necessary. Bell, after all, had time on its side; the more extensive the system, the more consumers would have a stake in its well-being. This was a lesson that the Diefenbaker Conservatives learned well. They also understood how hard it would be to initiate regulatory reform. In the meantime, Bell continued to ride a wave of demand for communications, and the board continued to sustain it.

REGULATION AND BELL CORPORATE
ORGANIZATION, 1960–1966

The decade of the sixties was generally salutary for Bell's business. The company continued to pursue its program of expansion and modernization, generally limiting its appearances before the board to questions of new stock issues or matters of interconnection. Still, problems persisted for the utility over the perennial question of its privileged relationship with its wholly-owned supplier, Northern Electric. This section will examine how both utility and agency handled the problems of Bell's corporate organization. While the issue did not have the salience of the regulatory problems of the 1957–9 period, existing problems and solutions offered by the board demonstrated the limits of federal telecommunications regulation.

While other sectors of the Canadian economy were suffering from recession, Canadian utilities posted profits in 1960. The *Financial Post* reported that the telephone utilities had done particularly well – so well in fact that Bell began building a multi-million dollar administrative centre in Toronto. The company's 1961 profit had jumped 16.9 per cent over the year before, sustained by the rising tide of consumer demand for telephones and what the company called "optional and special services." TWX, a teletype service, rapidial, wide area telephone service, facsimile, and the increasing demand for data communications were becoming lucrative parts of Bell operations. Customers' changing communications requirements provided Bell with an opportunity to expand the use and indispensability of its network.[60]

Bell did not apply for another general rate increase until the late 1960s. Instead, its appearances before the BTC were necessitated by minor revisions to its long distance tariffs, new stock issues, or legal entanglements with subscribers over service or interconnection. By and large, regulatory questions in the early 1960s did not possess the urgency of the previous ones. The changes approved by the board in the early 1960s generally confirmed the company's shift to modernization of its plant. Table 10 shows the national picture of modernization, measured by the use of automatic switchboards. The next table shows the increased traffic in both local and long distance that the switchboards had to handle.

Bell's relationship with Northern Electric had been an issue in every postwar rate case. The municipalities had held up the relationship as one major source of untapped revenue, especially since Northern's operations were neither regulated nor included in Bell's rate base. Indeed, Bell executives would seek to rehabilitate their subsidiary by removing it even further from regular Bell operations, to ensure that the company would wean itself from its dependence on Bell operations and learn research and marketing on its own.

Another attempt was made to open the question by one of Northern's competitors, Industrial Wire & Cable Company, in July 1963, and heard by the BTC in January of the next year. Industrial asked the board to restrain Bell from holding shares in Northern Electric. If that were not possible, then Northern's business should at least come under board scrutiny. Industrial's case was heavily based on the American regulatory experience. There the transactions between a regulated company and any affiliates were subject to examination, since the relationship between utility and supplier was deemed to be non-competitive. Accordingly, thirty-one states had extended regulatory jurisdiction to utility affiliates. Therefore Industrial argued that the BTC had a duty, if it wanted to ascertain the reasonableness of Bell's rates, to regulate its corporate affiliate. The price comparisons that Bell had provided as proof was not sufficient, since there was no arm's-length bargaining for Northern's supplies.[61]

For Bell, the relationship was essential for long-term planning and systems development. Manufacturing facilities could be geared to the introduction of new equipment without the annoyance of dealing with outside contractors. The integrated planning benefited both Bell and Northern, since Northern could use the Bell orders to establish a low unit cost for their supplies.

Therein lay the problem for Industrial. Northern had obtained an unnatural advantage over its competitors in the field by virtue of its unregulated relationship to Bell. The remedy was to bring Northern under the regulatory umbrella, thus equalizing the competitive conditions between Northern and other companies. If Northern profits on Bell business were excessive, then it would reduce the return needed from telephone operations. Of course, such a cost accounting would require a considerable expenditure of resources on the part of the board, and an expansion of the board's regulatory ambit.

In its judgment in July 1963, the Board held firm to its traditional handling of the Bell–Northern question. The relationship between the two had been discussed as early as 1921, but Bell's right to hold shares of Northern had never been challenged until Industrial's application. The board decided that since a telephone line existed between the Bell plant on Belmont Street in Montreal and Northern's principal manufacturing premises on Shearer Street, the 19,000 feet of Northern line to Bell was enough to meet the legal tests that denied Industrial's application. Industrial argued that the word "line" effectively meant an entire telephonic system. Commissioner A.S. Kirk concluded that Northern's activities beyond telephony were "within the scope of the powers accorded to the company by its Charter."[62] It is a fitting tribute to the power of regulation in this period that what turned out to be one of the most important relationship for the history of Canadian technology turned on whether Bell and Northern were linked by a telephone line. Had Northern come under direct federal regulation, the subsequent development of Canadian telecoms would have undoubtedly been very different.

The issue did not end there. In a confidential study of Bell–Northern links, the Combines Investigations Branch charged that Bell's telephone subscribers had been over-charged at least by the "amount equal to the part of the purchase price of Northern's shares which was paid for out of Bell's retained earnings." The study also noted that Bell had been careful to see that the dividends paid to itself from Northern profits were as high as what Bell was able to earn on its own assets.

Was Bell profiting at the expense of subscribers? To be sure, the possibility existed. But even if Northern had been included in Bell's rate base, the study conceded the situation might be the same. The Combines study suggested that if Northern were regulated, Bell would not care if it lost money or not. If the company did, it would reduce Bell earnings and justify higher telephone rates to maintain a fair return. As matters stood, however, the opposite was true: Northern earnings were twice those of Bell. The conclusion: that the "impact of spreading monopoly has therefore been neutral." In the supply market, there was no way of escaping Northern's commanding position. Sales to Bell meant important competitive advantages for Northern which reflected the benefits of monopoly, and not superior efficiency. In an open market, Northern would not have fared nearly as well as it had with a captive customer. Bell executives, especially Robert Scrivener, had already acknowledged this with their plans to operate Northern separately to ensure it would make its way in the world.

The question of Bell's relationship with Northern Electric demonstrated two things about the board's regulatory practices. First, it was not willing to act upon a contingency: as long as Northern's profits were similar to Bell's, the effects were neutral on the rate base. Undertaking a major change in industrial structure via the board was too radical a surgery to contemplate. This despite the Combines Branch view that the board was the best vehicle for spin-off of Bell's non-telephone business, since neither the branch nor the courts could do it with such finesse and authority.[63] Secondly, if vertical integration held dubious advantages for the ordinary telephone subscriber, its benefits to expanding the system were undeniable. Bell was convinced that Northern could mould its production to Bell's objectives better than the competitive market. Both were powerful reasons to leave well alone. The board's handling of the Bell–Northern relationship was not especially controversial. Although the hearings kept Bell's operations in the spotlight, Industriae did not manage to convince either the BTC or large segments of public opinion.

Of far greater concern to the board were Bell's impressive profits. The Commissioners noted that the 1950s were a period of "unprecedented and radical change" with the number of telephones doubling in nine years. In those same nine years the company's gross annual operating revenues had tripled to $328.8 million. The period from 1958 to 1964 was an equally remarkable one in the eyes of the board. Bell had begun to earn in excess of its permitted level of earnings of $2.43 per share set in 1950 and confirmed in 1958. In 1964 Bell reported a record profit of $177 million on $2.71 per share. How much profit should Bell be allowed? The board hoped a review of the company's permitted level of earnings would answer that question. Perhaps more important was the need for the BTC to reassert itself after a period as regulatory wallflower.

The BTC gave respondents eight months to prepare their cases. The hearings would come seven years after the two politically-charged rate hearings of 1958. The list of interveners had grown long, and ranged from the Canadian Federation of Mayors and Municipalities to the Communist Party of Canada.[64]

The main issues at the inquiry turned on what the company's permitted profit level should be, and also on precisely how to arrive at it. In the past, the board had set a ceiling on the share price. Since Bell had been earning more than its per share dividend, it asked that it be allowed a 7 per cent return on its total capital as the new yardstick for rate determination. The company wanted to be free to improve its earnings on its own, and to meet "changing circumstances and conditions" occasioned by the strength and variety of demand for communications services. Now it contemplated switching to a percentage of total capital. To prevent the hearings from getting out of control, the board prohibited discussion on rate structure.[65] General telephone rates between 1957 and 1966 had only infrequently managed to stay below increase in the general price index. Telephone rates from the late 1940s to the mid-1960s as a whole outstripped increases in the price index. But the statistics are ambivalent, since telephone rates outdistanced the price index from the base year of 1949. In any event, rates were but one dimension of the problem, and one that the board was not anxious to consider.

In the twenty-two days of hearings the BTC heard the equivalent of 3,900 pages of testimony. Clarence Jackson, president of the Canadian Section of the United Electrical, Radio, Machine Workers of America, appeared to argue that any higher profit level for Bell "can only come from a transfer of income out of the hands of the users of telephone service through higher charges or for the withholding by the company of savings in the cost of service which should be passed on to users." Technological improvements had not resulted in greater cost reductions either. Jackson lent his support to calls for the nationalization of the utility, although it was never a serious option for the state.[66]

Lovell Carroll would not be so intemperate in his argument as those calling for nationalization, but he did acknowledge that company and country were at the crossroads. He lamented the lack of regulation of Bell–Northern transactions, especially since Northern was allowed to earn such a high rate of return. Since Bell subscribers had paid $543 million to the company in 1964, they had a right to expect that every avenue for cost-saving would be examined. The main thrust of the municipal case was not in opposition to the percentage formula, but holding the line at 6.1 per cent return on overall capital. The municipalities feared that if the board acceded to Bell's 7 per cent request, it would necessarily result in higher tolls to make higher earnings possible.[67]

On the last day of the hearings, Carroll proposed the creation of a Canadian public utility ombudsman. The suggestion was a product of municipal fatigue in fighting rate cases. Carroll, after all, was the lone lawyer representing 105 municipalities in Ontario and Quebec as well as four municipal associations; felling the Goliath was a monumental task for the beleaguered counsel. The ombudsman's duties, Carroll told the board, would be "to enter all rate cases under the jurisdiction of the Parliament of Canada; defend the public interest; prepare evidence; and cross-examine witnesses" – something only the municipalities had done collectively. This would alleviate the burden on the municipalities, especially since a good portion of the hearings covered old ground.[68]

Bell's spending was running at over $200 million per year, and its outlay since 1945 had been more than the combined total spent on the Trans-Canada Highway, the Trans-Canada Pipeline, and the St Lawrence Seaway. Bell counsel and expert witness peppered the hearings with predictions of decline if the company was not given the opportunity to seek a higher return to attract investors. A telecommunications network, especially one so advanced as Canada's, did not come cheap. While the company gave priority to its urban subscribers who demanded sophisticated service, it also recognized that rural areas wanted basic quality service for its sparsely settled sections. Combined with its desire to introduce swifter and more efficient technological advances, Bell hammered home its argument that management needed a decent permitted level of earnings on the one hand, and the continuation of its relationship with Northern Electric on the other.[69]

In the end, the board decided upon an overall rate of return on Bell's total average capital of between 6.2 and 6.6 per cent. In all other areas, the board affirmed its previous decisions over debt-equity ratio, Northern Electric, and the Bell–AT&T service agreements. In all,

Bell could walk away from its 1965 review with satisfaction. The company had got more or less what it wanted: a decision to adopt a rate of return on total capital as the basis for considering its earnings. The set rate of return would ensure continued growth. The board also established a "utility-type" rate of return basis for regulation which would allow the company to increase earnings in proportion to the growth in total capital invested in operations.[70]

The Bell inquiry was in many respects symbolic of the transformation of regulation from its postwar position. The inquiry did not attract the attention the 1957–58 hearings had, nor did they generate the accompanying political conflict. Still, the inquiry had shown the limits of the possible within the existing regulatory regime. In its reluctance to tackle substantive issues, the BTC had become increasingly unable to respond to public concern about the regulation of Bell. The result diminished its effectiveness, and reduced regulation to an iron cage.

CONCLUSION

This examination of telecommunications regulation reveals the promise and the peril contained in the Canadian style of regulation. The lack of a regulatory machine gave the board wide latitude, so that the commissioners could dispense decisions favourable to the company's interests and system development. It had the virtue of removing the pressure of decision-making from the political sphere. It also promoted a systemic view of the telecommunications system not easily achieved with other regulatory regimes.

The process was vulnerable, however, when it was engaged too often. Adverse economic conditions created a climate of conflict that rendered the process vulnerable. Seen in that light, the board's structure and practice favoured network expansion at the expense of its own ability to mediate between the economic interests of the company and the interests of subscribers. If those conditions persisted, the regulatory agency could become unsafe at any speed of development. Instead of incorporating players outside the regulatory game, the board adhered to its traditional strategy of respecting a widely defined managerial autonomy, even at the expense of forfeiting a significant portion of its power.

The results were mixed. Board decisions provided a fair return on investment, guaranteed the company's long-term financial security, protected its manoeuvrability in the marketplace, protected its relationship with its supplier, and set rates which reflected to some degree a balance of residential and corporate interest. The BTC also presided over an extraordinary period of enlargement of the telecommunications network. The benefits were not hard to discern: commercial development, technological diffusion, economic prosperity. The credo based on that trinity would entrench itself in future telecommunications debates.

Regulation failed in two major respects in the postwar period. Once able to sustain mediation and legitimacy, regulatory forms acquired a rigidity that made them conspicuously less capable of performing those tasks. Judicial conceptions of equity that were an integral part of the regulatory culture were increasingly at odds with popular notions of fairness and common sense. By following the company's lead in embracing the technological consensus about telecommunications development, the board was also less able to respond to the concerns of subscribers or mediate between competing conceptions of technological progress and institutional structure. Both regulatory practice and the statutory limitation were causing atrophy. As a result, the board grew remote from the direction of the telecommunications, and from those who were affected by its outcomes.

The regulatory dynamics in this period also resulted in a weakening of the notion of community property, and did so in two ways. First, board decisions on rates, interconnection, and stock issues created a regulatory climate that fostered greater managerial autonomy. This did not mean that the agency was captured in the classic sense, but the public power was weakened. Second, the reasons behind the shift in the balance of power – the explosion of demand – began to redefine the subscriber as more consumer of a commodity and less subscriber to a public utility. Far from being a natural outcome of demand and technological change, this tendency grew out of Bell's development strategy and regulation's support of that strategy. By their policies, the BTC promoted that change, but were unprepared to deal with the demands that would result. This is a process that would not find its full expression until well after this period, but it is in the postwar that it took root.

In treating all rate hearings administratively, the BTC was in a sense a victim of its own isolation. By not acknowledging the policy-making implications of its rulings or even denying they existed, the board grew increasingly remote from the shaping of telecommunications, and from those outside the regulatory arena who were affected by its decisions. In the postwar period, then, telecommunications regulation had become an iron cage – unyielding, and confined to its narrow statutory and ideological limits.

By the mid-1960s the BTC was only partly able to defuse growing dissatisfaction with regulation. Ideologically, the board found it

difficult to respond to increasing calls for more sensitivity to consumer concerns. Lack of any political push for revamping its powers
further limited the possibilities for revitalization. What is more, the
legacy of the regulatory culture acted as a strong counterweight to
bold, effective action. Forbearance is often a virtue in regulation, but
by the mid-1960s it was indicative of the weakness, not the wisdom,
of how telecoms were managed by government. If policy-makers recognized the problem, they were reluctant to expend political capital
in redefining an institution which sufficiently diffused tensions and
deflected criticism from legislators. From the policy-maker's point of
view, the board provided adequate, if not ideal, regulation of the
industry. Although some recognized the need for reform, it would
take over a decade after diagnosis to treat the regulatory problem.

2 Connecting Canada
to the World: International
Telecommunications Policy,
1942–1965

Canadian telecommunications policy not only supervised Canadian connections at home; it also operated in an international context. The Second World War and its aftermath recast a number of familiar landmarks in both Canada's international relations and in the country's national affairs. Internationally, Canada's position in the world had been redefined in relation to Commonwealth and continent, while at home the combination of depression and wartime emergency had sired a more powerful bureaucracy interested in establishing a new economic and technological order. The Canadian state's management of international telecommunications interests reflected these developments.

Canada's response to the dramatic restructuring of international telecommunications illuminates how the government perceived the national interest in the context of postwar international politics and technological change. Policy-makers' reactions reflected Canada's anxieties about national sovereignty on the one hand and the varying speed and direction of technological change in communications on the other. This generation of public servants would exercise their impulse to plan, at the same time showing how overall state control over international telecommunications policy was piecemeal and incomplete. The tension between planning and "letting it ride" was to be the main circumstance determining telecommunications policy.

This chapter will examine several critical steps in the development of Canada's international telecommunications links: nationalization, the struggle to respond to increasing demand, the role of military communications systems in civilian communications, and the shifting international alliances that dissolved and reformed over satellite technology. Each of these international issues pressed reluctant governments towards a national telecommunications policy.

DEFENDING THE NATIONAL INTEREST: CANADA AND INTERNATIONAL TELECOM NEGOTIATIONS, 1942–1950

If the Canadian government was engaged in the prosecution of the final stages of the war by the autumn of 1944, it was also concerned with the how postwar arrangements would affect the national interest. The country's telecom links were considered to be of the greatest importance, and would have to be negotiated anew in a dramatically new global political order.[1]

But the desire to infuse order into the conduct of the international telecommunications business can also be seen as part of an increasingly confident public service now armed with greater authority and influence.

The plan to establish a range of international institutions to guarantee a new postwar order was a strong indication of the global desire for greater cooperation and coordination. In Canada, the bureaucracy created by the Second World War, and a more positive attitude about the role of the state in economic development, provided an important new context for government action in a range of sectors. International telecommunications was no exception.[2]

Canada's prewar international communications arrangements were hard-wired into the dominion's historic and political links with the Commonwealth. External telecommunications services were controlled by the decisions of the Imperial Wireless and Cable Conference of 1928. The conference recommendations essentially resulted in the formation of two companies. One company was a merger owning practically all the shares of the old cable companies and the Marconi Wireless Telegraph Company. The other, Cable and Wireless Limited, was an operating company which owned all the communications assets of the merger, cables previously owned by the governments of the Commonwealth. Together they ran the empire's overseas communications systems. Cable and Wireless also leased the wireless stations of the United Kingdom post office. At that time a Commonwealth advisory committee was set up which included representatives of the dominions and which was given certain powers of control over policy and rates. The committee was the official link between the various governments and the company. By 1938, rates between distant parts of the Commonwealth were drastically reduced and codified into the empire flat rate. Despite these reforms, the British empire's advantage in communications design and development had been squandered, lost to the Americans. Wartime had only accelerated the drift. In London, the Dominions Office was furthermore convinced that, with the excuse of wartime necessity, the Americans had developed an extensive external radio telegraph network which would undoubtedly compete with the empire's system after the war.[3]

The problem of reordering global telecommunications symbolized the ambiguity of Canada's general postwar situation. The war emphasized both the necessity of the American connection, and the importance of Commonwealth ties. As the senior dominion in the Commonwealth, Canada occupied a position of first rank in imperial councils. The Canadian opposition to calls for closer imperial ties in 1943, however, provides a good indication Canada's ambivalent thinking when the time came to reorganize Commonwealth telecommunications.[4] In order to understand Canada's position, it is necessary to know something about Commonwealth planning for the postwar.

As early as 1942, a Commonwealth Telegraph Conference had issued calls for a postwar wireless communications network linking together the various parts of the Commonwealth. For its advocates, there was little time to lose. Facsimile transmission was a classic example of how Commonwealth countries were not keeping up with developments in the field. The United States, on the other hand, had presided over a communications boom in facsimile and other types of transmission.

The need to present a common front produced various schemes for imperial solidarity and closer ties. But Australian and British pressures to cement those links in communications left Canada unmoved. In any event, Canadian officials felt any such plan ran counter to the "deepest instincts of the self-governing nations" and was ruled out as impossible to sell at home. Though cognizant of the value of imperial ties in wartime, the Canadians were increasingly motivated by the larger idea of a United Nations. The Canadian position can be summed up as yes to Commonwealth consultation, no to central coordination. This attitude would set the tone for Canadian policy, especially in upcoming discussions on communications issues.[5]

The Canadian government had good reason to move away from closer imperial ties. By the autumn of 1944, the Americans were beginning to contemplate the postwar configuration of telecommunications, especially with a view to settling some outstanding points. A major irritant had been the refusal of Cable and Wireless to allow direct radiotelegraph and radiotelephone circuits between the United States and points in the British Commonwealth. The Americans were also anxious to break the Empire preference and establish uniformly low telecommunications rates.[6] The British dominions secretary was open to discussion on these points, but was anxious to complete the arrangements and discussions within the Commonwealth before dealing with the Americans.

Cable and Wireless of the United Kingdom had operated Britain's external telecommunications systems since 1929. But depression and over-capitalization had seriously compromised the company's financial position in the 1930s, so the company turned to the Imperial Rates Conference of 1937 for relief. The conference approved a company plan for a rate structure which heavily favoured Commonwealth interests. The plan was two pronged: rates would bolster London's dominant position as hub for news and commercial messages; and the company would strive to block American firms from establishing direct circuits between parts of the Commonwealth and the United States.

Although Canadian–US traffic was not affected, the Canadian government was opposed to any trend to imperial concentration on two grounds: cabinet was anxious to avoid any hint of imperial monopoly in communications; and it was opposed to treating the Commonwealth as an organic unit in telecommunications, as in other fields. Canada hence refused to consider any collective responsibility for the maintenance of strategic cables, although care for those terminating in the country would be considered.

During wartime, direct radio circuits were established between the United States and many parts of the Commonwealth. Cable and Wireless reluctantly consented, on the understanding that the circuits were to be operated only for the duration of the war. The company also refused to consider a revision of the 1937 Empire preference that would have reduced the cost to American wartime traffic. The British representative on the Commonwealth Communications Council, Sir Campbell Stuart, was concerned by reports that American commercial wireless interests were lobbying the US State Department to secure concessions in the British Commonwealth before anything could be arranged.[7] US secretary of state, Cordell Hull, expressed the hope that a new international telecommunications organization would be established to promote the orderly development of communications networks. The largest telecommunications interests would naturally have more control.[8]

The chairman of Cable and Wireless, the redoubtable Sir Edward Wilshaw, saw nothing but American encroachment in imperial affairs. He felt that in any dealings with the Americans Britain should adopt a "*non possumus* attitude, despite the changed world situation and our very much closer relations in all fields with the USA due to the war alliance." Otherwise London would cease to be the world communications capital. The Commonwealth governments sought to mitigate potential damage to inter-allied relations by granting American requests for direct circuits at the Commonwealth Telegraph Conference of 1942. Since communications questions were becoming more insistent, a new Commonwealth Communications Council was created, with equal representation from each dominion.[9]

Wilshaw eventually found himself fighting a war on two fronts. American wireless interests were on one side, and increasingly, the dominion governments on the other. To Wilshaw, Commonwealth communications were strong when the company's bottom line was healthy. Not surprisingly, the governments of the Commonwealth viewed things differently. They had agreed by 1944 that Cable and Wireless had demonstrated "Eastern Telegraph/imperial thinking." The Empire Press Union and Reuters were moreover, "gravely dissatisfied" with the level of service. Worse still, Cable and Wireless had become an obstacle to the efficient coordination of the Commonwealth communications network. The Dominions Office warned that if nothing were done, the dominions would go their own way.[10]

Britain's apparent goals in all of this were simple. To regain the prestige of the Commonwealth in telecommunications, a comprehensive communications network had to be built. At the very least, London had to be preserved as the news distribution centre of the Commonwealth. The challenge from the Americans had either to be accepted, or acquiesced to. Recognition of that challenge, and exasperation with Cable and Wireless, led to a plan for the radical reorganization of telecommunications arrangements. The Anzac Scheme, proposed by Australia and New Zealand, sought to provide "functional unity while enhancing local autonomy."[11] The main objective of the plan was to dissolve Cable and Wireless and replace it with nationally owned public utilities companies administered by respective Commonwealth governments.

The Canadians favoured the plan. Unlike its partners, however, the federal government had to worry more about the Americans. The chasm between rates charged inside and outside the Commonwealth had long been an irritant between the two. But in 1944, the rate differential had become a highly political question in the United States. Both Republicans and Democrats had made freedom of communications part of their national platforms. Commonwealth communications reform, and especially the possible elimination of Cable and Wireless, would please the Americans. Their joy at this outcome would be tempered, however, by their lingering suspicions of an imperial monopoly in telecommunications. The Canadians were quick to allay their ally's consternation by strongly supporting the creation of a global communications authority after the war. In 1943 the British War Cabinet decided that the time was not yet right to tell the United States that the Commonwealth had decided in principle to maintain direct wireless circuits, since nothing had been formally decided.[12]

The Canadian Cabinet War Committee approved the principle of nationalization of its international telecommunications links in the autumn of 1944. The United Kingdom was not so keen, and sent Lord Reith of the BBC around to Commonwealth capitals to garner support for a more imperial-friendly plan that would give London greater control. Working with Wilshaw, Reith advocated the creation of a single communications company, centred in London. The Reith mission visited Australia, New Zealand, India, Southern Rhodesia, and South Africa, gathering support for the alternate plan. Canada viewed the support for the "Single Empire Corporation Plan" with some suspicion, since it looked a lot like a tarted-up version of Cable and Wireless. Attempts to dress up the plan to resemble the Anzac scheme only increased their suspicion. In a terse note to the British high commissioner, the under-secretary of state for external affairs, Norman Robertson, made clear that Canada would have no part of the plan, since Canada was not prepared to "subordinate a Canadian publicly-owned telecommunications corporation to … centralized control." Any pretensions to re-imperializing Canadian-Commonwealth relations would be met with resolve in Ottawa.[13]

More congenial to the Canadians was South Africa's proposal of a consultative body to replace the central corporation. The real urgency, according to Mackenzie King, lay in dealing with the Americans. Negotiations over rates and use of circuits had to happen "at the earliest possible date." King wrote that Canada would take no action that could jeopardize "our common interest in reaching, if possible, broader international agreements in this field, and in particular, in securing satisfactory measure of co-operation with the United States."[14]

As governments prepared for the 1945 Commonwealth conference on communications, it was evident that Canadian and South African demands for a loose arrangement would have to be met. Agreement was thus sought on the creation of a consultative body, the nationalization of Cable and Wireless's assets and the disposition of oceanic assets (transoceanic cables, ships, equipment, stations, and so on).

The problem of the Americans was resolved by offering "complete collaboration with the United States." That meant the removal of points of friction between the Commonwealth and the United States, especially in the areas of rate-cutting and differential rates.

The Canadian cabinet discovered that, even taking into account Canadian and South African objections, the British corporation to be set up would own all oceanic assets and hence dominate the proposed central body. The Canadians then argued that the new agency should have centralized control over oceanic assets, especially since "experts have frequently argued that the system should be considered as a whole and that cable and wireless networks cannot be operated as efficiently independently."[15]

There was another problem. The proposal for a British company to operate the system with Commonwealth approval turned out to be merely a version of the old system. That situation produced one antagonistic company (Cable and Wireless), one exasperated major power (the United States), and one great strain on us–Commonwealth relations. From that point of view alone, any change in management would be better for relations between the Commonwealth and the United States. Canada's Interdepartmental Committee on Telecommunications Policy (ICTP) recommended that the country pursue a policy which supported the creation of a central body, and that oceanic assets be administered on a regional basis. This would give Ottawa the freedom to negotiate on telecommunications matters with any other country. The Canadians were willing to contemplate complete withdrawal from any proposed scheme if these conditions were not met, a neat symbol of Canadian deference to American telecommunications interests.[16]

The growing momentum for nationalization prompted Cable and Wireless to try to block any plan for the expropriation of their assets. The company argued that the Commonwealth nationalization scheme was a slippery slope; it might also inspire other governments to nationalize the company's foreign concessions. Wilshaw tried to pitch a last-ditch scheme to strengthen unity in Commonwealth communications while preserving the existing set-up, and where foreign interests could eventually participate. Wilshaw argued to anybody who would listen that the proposed new scheme would perpetuate the deficiencies of the League of Nations by ensuring that one voice could upset any potential agreement. In an act of desperation, the company even relented on the issue that had so rankled the Americans – differential rates. The company even softened its stand on use of direct circuits, something that its previously considered too suicidal to contemplate.[17]

Canadian reaction to the company's new proposals indicates how little agreement existed within the federal government over this issue. The minister of reconstruction, C.D. Howe, had been disturbed by the seeming rush to nationalization that had characterized the attitude of certain Commonwealth governments. The company's new proposal gave Howe the opening he needed to take ICTP chairman Evan Gill to task over the committee's strong support for nationalization. Howe argued that the Reith mission proposals would weaken the existing system of Commonwealth telecommunications without any visible benefit. Cable and Wireless, on the other hand, had put forward proposals that "commend themselves to the Government of Canada." Contrary to ICTP thinking, Howe argued that Canada had been well served by Cable and Wireless and its Canadian associate, the Canadian Marconi Company, and saw no compelling reason to change the status quo. Howe also believed that Britain itself, and not the company, was the source of Cable and Wireless's disagreeable relations with the United States. Britain had consistently refused to enter into discussions for fear of irritating the Americans, and "recent actions detrimental to the operating position of Cable and Wireless [were] the outcome of that feeling." The minister must have known of the dissent in Britain as well, where Lord Beaverbrook had serious reservations about the nationalization scheme.

Howe eventually acquiesced in the nationalization of the communications business of Canadian Marconi, but only reluctantly. But this was as far as the government was prepared to go: "Canada did not wish to be responsible for the management of the Telecommunications System as a whole."[18] The suggestion that Canada own its Pacific and Atlantic cables was also out of the question.

There was obviously disagreement as to how best to pursue Canadian interests in telecommunications policy at war's end. The ICTP, under the chairmanship of both Norman Robertson and Evan Gill, had come to believe that a political solution was appropriate, while Howe and his supporters in the Cabinet War Committee were much more reticent about such a plan. Howe's stake in the matter was perhaps tied up with fostering closer continental links; the nationalization plan would have meant closer financial and operational cooperation with Commonwealth governments. Cabinet reluctance was also indicative of the serious reluctance to set aside market forces.

Only some of Howe's objections found their way to the Canadian delegates to the Commonwealth Telecommunications Conference held in July and August of 1945. In the interest of greater Commonwealth relations, cabinet agreed to create a public company to operate external communications facilities within Canada. It was also prepared to cooperate in the creation of a central advisory body. On the question of oceanic assets, the instructions were clear: Canada wanted them to remain in private hands.

At the conference, the plan to administer oceanic assets via a central agency was quickly dropped when both Canada and South Africa indicated they would not participate. As instructed by Howe, the Canadian delegation stressed their satisfaction with the company and Canadian Marconi. The deciding factor against Cable and Wireless even for the favourably disposed Canadian delegation was the problem of Commonwealth–us relations. As the private report of the delegation took pains to stress, "The Companies proposals did not offer any material prospect for improving relations with the United States with whom their policies in the past had caused resentment."[19]

The conference settled several vital issues for the future of external telecommunications. Consensus was reached on the nationalization of the operating companies of Cable and Wireless, but not its holding company. To that end, the conference worked out the details of the revenue-sharing agreements between the partner governments. Canadian reservations were based both on the proposed financial deal and on the timing of the nationalization process. The financial arrangements included a volume-based revenue pool. Nationalization would have to await the outcome of the negotiations with Cable and Wireless in Britain, as well as the outcome of discussions with Canadian Marconi. Given the good relationship between Canadian Marconi and the federal government, there was every hope that negotiations could proceed at a leisurely pace, and with full respect for the company's interests.[20] Besides, Marconi was busy developing other lines of business.

The conference also agreed to expedite an agreement with the United States over outstanding telecommunications issues. In the interests of good relations with its neighbour, the Canadian government was asked to consider ending its preferential rate for government telegrams in order to end American resentment at "discrimination." The government was also asked to increase the revenues of Cable and Wireless, to be applied to a lower new world flat rate that would effectively reduce the gap between the Commonwealth rate and the rest of the world.

If the organization of telecommunications was important from a Commonwealth point of view, the settling of outstanding telecommunications issues with the United States was absolutely vital. While Canada did not have as much at stake as other Commonwealth governments owing to separate agreements, an amicable settlement over rates and curbing unwanted, uncontrolled competition was highly desirable.

The main irritant for the United States was the Empire preference. Most Commonwealth governments were prepared to reduce the preference to a "20 per cent tolerance," but getting rid of it entirely was out of the question. Canada, on the other hand, was prepared to eliminate the preference entirely or extend it to other countries. The same would apply for the differentials in the press rate.[21] On the question of direct circuits that had been established in wartime emergency, Canada proposed that after consultation with partner governments, any government wishing to do so should be able to open any new direct circuit, with the proviso that the potentially cost-saving routing of transit traffic be "resisted as far as possible."[23] The balancing act between accommodation and self-interest always required artful compromise.

Another conference, this time in Bermuda, was held in December of 1945 with delegations from every Commonwealth country and the United States. Tellingly, the US contingent included seventeen observers from American corporations, while the Commonwealth companies were fewer, but included Canadian Marconi, Canadian National Telegraphs, and Canadian Pacific Telegraphs. Compromises were reached on the use of direct circuits, which limited their use to Commonwealth traffic. This came as very good news to Commonwealth governments anxious over the financial viability of the organization's cable systems. Agreement on rate-cutting was also achieved, to the great relief of the Commonwealth delegations.[23] In return, the British conceded to American companies the right to establish direct circuits between the United States and points in the Commonwealth, especially where Cable and Wireless exercised a monopoly.

From Canada's point of view, after the Bermuda conference, "there remain[ed] no outstanding unsolved issues in the field of telecommunications between the United States and members of the Commonwealth." Outstanding issues in telecommunications had been a significant obstacle in Commonwealth–American relations for fifteen years. The Empire preference was maintained, "uneconomic" rate-cutting was eliminated, and the danger of American or British unilateral action was contained. The Americans came away with some notable victories, most notably through the removal of Cable and Wireless. Not least was the noticeable improvement in Anglo-American relations. Although Canada gained little in terms of rate reductions, it was allowed to retain the government preference for its own telegraph messages. Canada could also look forward to increase routing traffic for press messages.[24]

The four years after the war's end also settled other important questions in communications, most notably the International Telecommunications Union Conference of 1947, the International High Frequency Conference in 1948, and the International Conference on High Frequency Broadcasting in 1950. However significant these conferences were in determining matters such as spectrum allocation and standardization, the Bermuda conference clearly held the most import for the configuration of Canadian telecommunications.

The Commonwealth Telecommunications Board (CTB) was incorporated in the United Kingdom on 31 May 1949 by the Commonwealth Telegraphs Act. The board would regulate the Commonwealth's external wireless and cable telecommunications. As discussed above, the Commonwealth agreements called for the application of a specific rate and revenue pool formula and provisions for overall maintenance of the Commonwealth telecommunications system.[25] The board also promoted communications research, although the British National Body supported the lion's share of the basic research.

Canadian domestic response to the Overseas Telecommunication Act in late 1949 is revealing. To the CCF, the operation seemed to uncover a salvage operation for Canadian Marconi. Since the war, the company had posted substantial losses ($115,000 in 1946 alone), and its current assets had dropped by almost half a million dollars by 1949. "Why is it," Ross Thatcher of the CCF asked, "that every time a company gets into difficulties the government has to pull it out?" Other MPs viewed the establishment of the Canadian Overseas Telecommunication Corporation (COTC) as further proof of the regrettable estrangement of England's former colony from the imperial centre. Perhaps the most telling comment came from Donald Fleming of the Conservatives. After scolding the CCF for not supporting or at least having reservations over the bill because of nationalization, he added: "It seems to me this is a field which is often regarded as the *natural* object for nationalization, namely, a public utility and yet the socialists are warning us against nationalizing this company because it has none of these wicked profits." Generally, the government encountered little opposition to nationalization. The arrangement was seen as a natural outcome, especially for a utility-based enterprise like telecommunications.[26]

Nationalization negotiations were not completed until three years later, in the spring of 1952. Canadian Marconi had viewed the Commonwealth decision to nationalize external telecommunications with frustration. The executives at Marconi felt pressure from their parent Cable and Wireless to obtain advances to repay outstanding loans and investments, while at the same time to get as much as they could for the assets.

That was easier said than done. Canadian Marconi's good standing with C.D. Howe and an earlier Canadian cabinet did not translate to an easy ride in negotiations with D.F. Bowie, president of the COTC. The parties met every month from 1950 to 1952 to reach a deal. As time passed, Canadian Marconi's asking price tumbled from $3.4 million in 1950 to the final price of $2.7 million in 1952. Although the company's president, S.M. Finlayson, believed anything over $2.5 million would represent "a very handsome capital gain," he resented COTC's hard bargaining. The company also felt that Bowie and the COTC were taking advantage of the "political aspects of the matters" and intentionally delaying a deal to pressure the company. COTC's continuous threats to take the matter to the Exchequer Court would almost certainly have been disastrous for Marconi, since any settlement might be based on the company's weak earnings record. With the spectre of a ruinous judgment firmly in their heads, Canadian Marconi settled in May of 1952, and Bowie and the COTC got their company.[27]

The new crown corporation had been capitalized at $4.5 million at incorporation amid doubts about the financial solvency of the company it had just nationalized. The COTC began operations with overseas facilities that amounted to three telephone and thirteen telegraph circuits provided by high frequency (HF) radio and telegraph cables. It was only in 1963 that the COTC's mandate was extended to provide communications with non-Commonwealth countries, although it had been doing so for some time.[28]

COTC's external telecommunications services received a significant boost in 1953 when the federal government began discussions with Britain and the United States over a trans-Atlantic telephone cable, to supersede the telephone communications between Europe and North America, which was then conducted by problem-ridden high frequency radio. Typically, the problem was who would control it, and who would use it? Canadian interests found themselves once again competing for position between US commercial and British national interests, and was the weakest point in the triangle.

In discussions with the British, the Canadians secured the right to exclusive use of up to five telephone circuits between Canada and the United Kingdom. All transoceanic telephone calls originating in Canada were to be handled by the COTC. American Telephone and Telegraph (AT&T) insisted on giving their Canadian subsidiary the right to install, own, and maintain operating equipment as well as the land cable to the American border. If the Americans were permitted to operate on Canadian territory, AT&T would displace the COTC in trans-Atlantic communications.

Cabinet agreed that there were sufficient advantages to going it alone, should AT&T not back away from their demands. The gambit was risky, since AT&T had threatened to construct a trans-Atlantic cable via the Azores. Further discussions were held throughout the summer of 1953.[29] But after three months of laborious negotiations, AT&T was still adamant: their subsidiary would have to control the Canadian connection of the cable, and the parent company would have to own 50 per cent of the cable itself. Exasperated, the Canadian government began to consider a two-way deal between Britain and Canada, cutting AT&T out altogether, even though the plan would come at an additional cost of $10 million. The Canadians would ask the British to consider whether they really wanted "an American company maintaining and operating a portion of the Commonwealth Telecommunications network," and hope for the right answer.[30]

The debate in cabinet focused on the principles at stake: sovereignty or private enterprise. Those who argued for acquiescence in the AT&T proposal suggested cost was paramount, and that the political and security dimensions were not significant. Nationalists in the cabinet, however, argued that the political and security considerations were real enough, especially since the scheme involved American control of Canadian telephone and telegraph facilities.

The cabinet debate took another turn when the British Post Office refused outright any Canadian overtures for a bilateral deal. Commonwealth unity was one thing; but this was business. The cabinet wavered on the AT&T deal, since Canada could pay for 9 per cent of the outlay and get six to nine circuits. The only alternative was to continue discussions with Britain and position the COTC to exploit all Canadian experience and knowledge in cable and wireless technology in order to come back to the negotiations prepared to defend Canadian interests effectively.

Negotiations were concluded in October 1953, but only after the COTC was able to secure exclusive Canada–United Kingdom use of six telephone and six telegraph circuits. AT&T relented on the issue of the rights of its subsidiary, and agreed to shelve plans for operating a microwave relay system in Canada, save for transmission purposes. The only thing that saved the deal was in the fine print of a 1927 agreement that AT&T had concluded with the secretary of state which had allowed the American communications company to transmit inside dominion territory, but prevented further expansion of a relay system for other than their own cable requirements. The trans-Atlantic telephone cable agreements were approved by order-in-council in mid-October 1953. The trans-atlantic cable, TAT-1, was put into operation in September 1956 amid great public interest.[31]

The negotiations over international telecommunications sparked the interest of Norman Robertson. From his post as high commissioner for Canada in the United Kingdom, he argued that Canadian policy should be mindful of the pitfalls of participation. For one, corporate and national investments in less efficient means of communications were hindering the diffusion of cheap and efficient trans-Atlantic telecommunications. Robertson hoped to use cheaper and faster communications links to put some life into article 2 of the NATO treaty, which promoted economic and cultural ties.[32] As matters stood, there was no authority to promote and coordinate those links. He suggested Canadian policy be prepared for quick, cheap, and dependable intercontinental communications in five years' time. Regulation would have to follow.

In the meantime, the COTC opened direct circuits with Germany in 1955, direct radiotelegraph service with France in 1956, and with Rome and Tokyo in 1958. In September 1958, responding to technological advances in cable quality, microwave ground relay, and submarine cable technology, talks began for the implementation of a Commonwealth "Round the World" system. The plan was prompted by fears that the Commonwealth's position in global telecommunications would be lost piecemeal to foreign competition.[33]

The gradual expansion of the COTC in an unfolding market for external telecommunications services did not go unopposed, especially by businesses that had been providing external communications before the creation of the crown corporation. As well, the transportation and communications sector represented a significant portion of economic activity in Canada. In the mid-1950s the COTC bought the Commercial Cable Company, a move that provoked considerable opposition in Parliament and the private sector. The Conservatives had previously argued for the existence of a strong COTC; in 1956 their attention had turned from a strong Canadian presence in Commonwealth telecommunications to the possible costs to the consumer of having a growing monopoly in telecommunications. "What this Parliament should be endeavouring to do," MP George Hees argued, "is get the best possible deal for the consuming public of Canada, and that can only be done by enabling private corporations who wish to enter the field and take a chance of making a success in that particular line of business." The CCF saw no compelling reason to nationalize, and opposed the COTC's expansion at every turn.[34]

George Marler, the minister of transport, defended COTC plans on technological grounds. The introduction of coaxial cable had "completely revolutionized" telecommunications and especially telecommunications by submarine cable. But coaxial cable could only carry so many circuits, and there were good economic grounds for licensing a limited number of cables. The refusal of the federal government to grant an overseas licence to Commercial Cable stemmed partly from these reasons. The ultimate reason, as Marler explained, was protectionist. The Commercial Cable application signalled that "Canadian telecommunications operations are now snowed under in a field which is highly competitive" and in which large US corporations could command very substantial sums of capital to outbid Canadian enterprise.

Marler's contentions were justified. By the mid-1950s the explosion in traffic demand for commercial communications rendered existing plant inadequate. The next year, the British Post Office, the COTC, and Cable and Wireless planned another cross-ocean link in CANTAT. The undertaking represented a technological advance based on British lightweight cable and repeater technology. Through its 2,100-mile route, CANTAT achieved 70 per cent greater circuit capacity than TAT-1 could achieve. CANTAT also provided the prototype for a greater effort. The Commonwealth's recognition of the technological gap between demand and facilities resulted in a conference held in London in the early summer of 1958. Partner governments approved plans for a new Commonwealth "Round the World" cable infrastructure whose costs would be borne equally. The second section linking Canada with Australia and New Zealand (COMPAC) and South East Asia (SEACOM) was completed in early 1964. The COTC would own the cable from Vancouver to south of Oahu in the Hawaiian Islands. On the Atlantic side, ICECAN would connect Newfoundland, Greenland, and Iceland, linking the system with SCOTICE at a cost of $3.5 million.

CONVERTING MILITARY SYSTEMS TO CIVILIAN USE

Another problem that confronted Canadian policy-makers in the postwar period was the extension of the US–Canada military communications systems and their conversion to commercial civilian use. In this field, Canadian public policy was barely equal to the task of managing important questions in telecommunications, which eventually led to calls for a more coherent national policy.

Between 1945 and 1960, the state responded to problems in telecommunications policy as they arose, principally in international and military matters. The centralization and coordination of national administration that flourished in wartime emergency continued in international telecommunications policy in war's aftermath. But it is instructive that the apparent success of the idea of central planning was arrived at reluctantly and incompletely. Dispute over the extent of state control and outright opposition to an expanded supervisory role for the federal government prevented the full flowering of central planning. As a result, policy-making in international telecommunications policy was conducted ad hoc, and despite suspicions about its efficacy.

The centre of any future hopes for a more articulate telecommunications policy lay in the Interdepartmental Committee on Telecommunication Policy (ICTP), an umbrella group set up to oversee and coordinate telecommunications policy discussions. The committee consisted of representatives from most departments and was originally under the direction of H.O. Moran at External Affairs. Although telecommunications matters were considered important enough to merit a policy committee at the interdepartmental level, Moran noted that "it was not easy to devote to this Committee, which at that time was extremely active, the time which its matters required." Moran's successor was also involved in the ICTP, but A.F.W. Plumptre was unfamiliar with telecommunications problems and also had difficulty giving the committee his close attention. Because DEA officials were neglecting the committee, the chairmanship fell to the controller of radio in the Department of Transport, G.C.W. Brown.[35]

Under Brown's direction, the committee became much less active through a combination of benign neglect and the feeling at DOT that the major issues – the establishment of the Overseas Telecommunication Corporation and the preparation of a Canadian position at the International Telecommunications Union conferences – had been disposed of. A Joint Telecommunications Committee memorandum suggests an additional reason why interdepartmental policy-making was treated with such indifference:

The Department of Transport not only represents the majority of Canadian civil users, but is also responsible in Canada for the control and enforcement of national and international radio regulations. The DOT is, therefore, the proper agency to coordinate and supervise the technical working in this field.[36]

The attitude that the DOT was the proper agency would resurface occasionally when its autonomy in telecommunications was threatened. But, as George deT. Glazebrook at External Affairs noted in

1952, "there appears to be an absence of any governmental authority on general communications questions."[37] Ironically, disagreements about who should run telecommunications policy were preventing any policy at all from being articulated.

The committee's postwar agenda was filled with matters of mostly military concern. After the specifics of nationalization of civilian communications had been dispensed with in 1949, the committee concentrated on radio problems. Dependence on radio by civilian and military agencies had overburdened the frequencies allotted to Canada and needed close scrutiny and action. In a document entitled "The International Control of Radio Frequency Assignments" in May of 1950, the JTC chair, Group Captain E.A.D. Hutton, explained the importance of telecommunications policy: "The vital economic, political and military interests of Canada require that immediate action be taken to prepare a civil/military policy on the international frequency assignment affairs, and that, commensurate with our resources, the strongest possible Canadian civil/military representation be provided at future international conferences on this subject."[38]

The coordination of these and other important telecommunications matters had been dealt with by the occasional liaison between departments. This was the machinery in place to handle the "tremendous, war-accelerated expansion of telecommunications." The complexities of radio frequency allocation required resources and expertise, and ICTP secretary Evan Gill hoped that the committee could be given a more general policy role. More and more, in his view, Canadian interests abroad had suffered at international military and civilian discussions on telecommunications matters because of lack of policy coordination. The Department of Finance strongly supported the need for greater coordination, especially since international financial arrangements and government traffic were ongoing concerns. Gill's attempts to fully reactivate the ICTP in 1950 failed, mainly because of opposition from DOT. The department jealously guarded its jurisdiction, perhaps even at the expense of the formation of a new policy. But what was once sufficient was so no longer.

Canada–US military relations heavily influenced discussions about telecommunications policy through the 1950s. Nowhere was this more evident than at the Privy Council Office, There yet another committee – the Ad Hoc Committee of the Working Group on Telecommunications Policy – had been established in the spring of 1955. Debate over military communications systems had sparked discussion about the nature of Canadian communications sovereignty. Military decisions about communications operations, and the relationship of these facilities to the United States and other countries, were taken on a piecemeal basis. When military imperatives no longer dictated the need, American military systems in Canada could be converted to civilian use, and that would generate worrisome problems. Who, for example, would own them?

Whether that military necessity would ever dwindle was open to question. Lieutenant-Commander A.A. Beveridge noted that he did not foresee the military imperative ever shrinking "to pre-1947 dimensions." The Arctic had become a permanent frontier in defence terms. The Distant Early Warning (DEW) Line was here to stay. The principal concern was not about the necessity of the DEW line, but whether the US military would establish squatters' rights. The working group noted that "unless we specifically protect against this contingency it might be difficult to recover the use of frequencies for our own civilian or other uses in peacetime which have been used by the military over a period of time." On the other hand, a refusal to allow the system to go ahead would cost the Canadian government $15 million, payable to the US airforce. "Despite our legal right to refuse permission to operate the [tropospheric communications] scatter system," the committee noted, "it may be difficult to avoid ill feelings and considerable expense, should Canada wish to recover these television frequencies at some future date."[39]

The Pinetree Agreement of 1951 governing US military activities on Canadian soil provided for mutual consultation on the establishment and removal of radar operations. The problem was conversion of military communications to civilian commercial use. Specifically, tropospheric circuits were being expanded from Frobisher Bay which would provide two international circuits linking the United States with Europe. "If the Labrador portion of the system is permitted to come under US control," the committee warned, "Canada may be unable to avoid exploitation by US commercial interests, both in peace and war."

In the end, technological advances saved Canadian policy from itself. The problem of military relations would continue until the mid-1960s, when the military communications infrastructure was less controversial, and the proliferation of commercial/civilian systems precluded the need to rely on convertible military systems.

Canada's experience with inter-allied military relations reinforced the state's idea that telecommunications and national sovereignty were closely linked. Events had also underscored the inadequacy of the existing apparatus of the state to deal with the important policy questions arising from telecommunications matters. Nonetheless, the postwar lessons learned in dealing with the Americans and British would be applied to safeguarding the state's interest in both international and domestic telecommunications.

REORIENTING CANADA'S INTERNATIONAL TELECOMMUNICATIONS POLICY IN THE 1960S

Canada's telecom link with the US military was not the only relationship pushing policy-makers to stake out the country's interests. Commonwealth institutions grew increasingly inadequate for the needs of Canadian communications, particularly in the face of satellite technologies. Indeed, arrival of the satellite favoured continental and other non-Commonwealth links.

Canada's Commonwealth partnership in overseas public communications development originally flowed from its former imperial links and its self-interest. Postwar arrangements allowed Canada to expand direct circuits around the world, and participate in a strong global organization. As one DOT aide-memoire explained, Commonwealth links enabled Canada to "have a greater voice and to maintain a more independent position in international public communications matters" than would have been the case if the country had relied solely upon the existing AT&T-dominated US overseas network. The protection of Commonwealth and international organizations thus gave Canada manœuvring room negotiators would not have had with exclusive American association. As early as 1957, however, the government began to contemplate the question of Canada's continued participation in the Commonwealth communications system.

Routing messages through outer space via Commonwealth channels offered some advantages, not least of which was greater Canadian control. But Commonwealth and European technology in the field of space communications lagged behind that of the United States. Acceptance of participation in any more advanced American system would mean accessibility as early as 1964 with operating systems in place in Europe, Asia, Australia, South America and the United States.

By early 1961 there was a growing recognition that at least two competing systems were going to be developed for political as much as econo-technical reasons. Canada was caught in a familiar position, having to choose association with one system or the other, or try to keep a grip on both. The major concern was to prevent market competition if two rival systems emerged. Competition, if allowed at all, would be limited to a "degree of rivalry" in service provision and quality. "It can never mean keen competition to the point of rate wars," DOT officials argued, "because this could lead to a deterioration in service brought about by a lack of interconnecting arrangements among rival factions."[40]

Britain was anxious to preserve its position in global communications, and was pressing Commonwealth governments to fall into line. The DOT argued that in spite of some problems involving the United States and Europe, Canada should participate in any system that would safeguard its interests in communications. Participation in a Commonwealth scheme also had the advantage of economy: capital investment in a Commonwealth/European system meant the provision of ground stations, a pro-rated share of development, and control, launching, and satellite facilities on the basis of the channels that would become available to Canada.

The question of financial viability of a civilian satellite communications system and its integration into the hard-wire network was a complex one. Canada's overseas requirement had been assessed by the 1958 Commonwealth Communications conference. The Commonwealth's cable system would have fulfilled the country's external communications needs. But the departure of South Africa from the Commonwealth in 1961 caused policy-makers to worry whether the UK–South Africa segment of the system would ever be completed, not to mention the spur from that cable into South America. All of this would result in a shortage of capacity.

With the advent of commercial satellite technology, the gulf between Canadian and Commonwealth interests widened. By 1965, the year that the Early Bird Type HS 303 satellite was launched to service the United States and parts of Europe, the Commonwealth Telecommunications conference was unable to reach any conclusions about the desirability of a Commonwealth satellite system, and referred the matter for greater study. The Telecommunications conference of the following year indicated that the carefully negotiated Commonwealth postwar arrangements arrangements had been overtaken by communications technology and explosion in demand.[41] Developments in satellite technology were revolving around stronger political blocs: the United States, the Soviet Union, and the European Community. This fact was recognized when the 1966 Commonwealth conference proposed the dissolution of the CTB. The board was dissolved in 1969, almost twenty years after its inception. Paradoxically, greater global integration meant leaving Commonwealth ties behind.

Canada's potential participation in the US satellite proposal was more problematic. While the Americans announced that all countries would have equal and non-discriminatory access, Canada's problem was over the question of ground stations. The government feared that the Americans would propose that Canada's satellite needs could be met routing Canadian traffic via New York City. This was not the first time this solution had been proposed: in the negotiations for the Trans-atlantic cable in 1952, a similar arrangement was proposed and declined. Instead, Canada opted for part ownership of the cable and the subsequent development of a separate British-Canadian cable. The preservation of some autonomy in the communications field was now at issue in space communications, and the DOT strongly felt that the old policy should be extended to the new technology. That would at least ensure the possibility of an indigenous technology base and a greater voice in the international system.

By the early 1960s, the COTC was considered to be of increasing importance from both military and civilian viewpoints. Indeed, telecommunications traffic had exploded by the early 1960s and telephone rates were getting cheaper as well, as technological improvements facilitated falling costs and additional capacity.

The COTC came to be viewed as a positive governing instrument – one that could potentially pilot the Canadian telecommunications industry to international leadership. "Besides leadership in Commonwealth telecommunications," one report on the crown corporation noted, "COTC could also play a dominant role in world communications in view of the special political, technical, and economic status which Canada enjoys."[42] Some policy-makers hoped the COTC might act as a spearhead for the Canadian telecommunications and electronics industry. The Glassco Commission on Government Organization envisaged the COTC acting in commercial and technical cooperation with the domestic telecommunications industry and promoting integration with both domestic and foreign networks.

The COTC had long provided the communications industry with needed business, facilitating more and more international traffic through private circuits. In the case of CN/CP Telecommunications, the corporation was viewed as an instrument to balance competition between it and the other common carriers. The railways were determined to capture the COTC's 1964 business, which would include thirty-six high-quality voice circuits across the country to connect the trans-Atlantic cable to the Pacific cable.[43] Increases in traffic would reinforce the necessity for expanded capacity.

As COTC's telecommunications activities became increasingly important, attention began to focus on the state's ability to remain in control of developments in telecommunications. The Glassco Commission reported that although nominally under the Department of Transport, there appeared to be no fully accountable and technically competent authority to which the COTC executive might appeal for advice and support on matters of further capital investment or system expansion. This masked a greater dilemma for policy-makers: the fundamental problem of state enterprise. The emergence of semi-autonomous crown companies – the COTC and Canadian National Telegraphs (CNT) in particular – operated in competition with private common carriers. As the industry developed and became more competitive, the federal government had to come to terms with the need for a higher level of coordination for telecommunications activities that would see those activities dealt with separately from the DOT.

The COTC receded from public prominence by the mid-1960s, eclipsed by the focus on satellite development. But the crown corporation's position in the telecommunications infrastructure was critical. Incremental technological change characterized the corporation's activities, along with an ever-increasing demand for international communications, had reached a point where the federal government felt it less necessary to oversee the corporation's activities closely. In addition, satellite technology prompted the Canadian state into the recognition of the inadequacy of Commonwealth institutions, and the ascending importance of American and other non-Commowealth links.

CONCLUSION

Several major conclusions result from the study of Canada's experience in international telecommunications. First, national sovereignty motivated Canadian policy more than consumer choice, or any other consideration. The aims of Canadian telecommunications policy in this period coincided with the country's struggle to reconcile national interests with the postwar reorientation of international relationships. Apart from rate regulation, state involvement in telecommunications policy as such meant the representation and defence of Canada's international interests. Telecommunications policy was concerned with staking out the interests of the Canadian system amidst the growing interdependence of states. The international concerns of national policy-makers reinforced the historic treatment of the Canadian telecommunications architecture as a single national system. This was a world where market forces were subordinated to political ones.

Canadian policy also had to find its place in newly formed international political arrangements. The ascendancy of the United States stood in contrast to the declining importance of the institutions of Empire and Commonwealth as the hub of the Commonwealth communications infrastructure by the mid-1960s. Still, Canada exploited membership in international organs such as the International Tele-communications Union and the Commonwealth Telecommunications Board as counter–ballast to the power of the United States. In all, the changing organization of telecommunications reflected an important change in state relationships in the North Atlantic.

Secondly, Canada's attempts to shape the direction and pace of international telecommunications policy seemingly indicates the triumph of central planning. Upon closer examination, however, the range and reach of government policy in telecommunications was far more ambiguous. After an initial burst of postwar energy, Canadian telecommunications policy-makers dealt with matters without a policy framework, and very much on an ad hoc basis.

Thirdly, the Canadian state perceived changes in communications through the optic of certain powerful historical models. Questions of international law and order in communications matters were filtered through assumptions about public policy and technological change generated by past experience. How could it have been otherwise? Canadian policy makers consciously or unconsciously bore witness to continuity by representing the state's historic concerns.

Also part of this inheritance is the ambiguous nature of state participation, often instigated by governments professing a commitment to the principles of free market economy. This legacy of a "mid-Atlantic" pattern of regulation and state enterprise evolved less out of an ideological commitment than out of Canada's position on the economic and political periphery of North America. This climate created the need for the positive state. Development of canals, roads, ports, railroads, post, telephone, telegraph and radio was thought vital to the developmental and integrative mission of the state. These traditions mattered when it came time to develop a national policy in international communications.

Lastly, experience of dealing with the military and international management of telecommunications led to pressures to overhaul the inadequate and ad hoc ways *both* domestic and international telecommunications policy were arrived at. The increasing difficulty of managing technological complexity contained within that experience, especially by the late 1950s, also provided significant impetus for change. Pressures for administrative renewal and rationalization combined with technological change were symbiotic, and an insistent reminder that as time marched on, federal policy was falling seriously behind.

3 Canadians and Computers: Initial Canadian Responses to the Computer, 1948–1968

The introduction and diffusion of new technologies involves a process of adaptation to local contexts. Technology shapes, and is shaped. This chapter will explore aspects of the diffusion of computer technology in Canada. Computers are linked by wires, connected to large databases and utilize telecommunications links to communicate with each other. More and more, they became an integral part of the communications infrastructure. The nature of the computer revolution in Canada, at least in its early phases, can be summarized by its rapidity of diffusion, its essentially imported nature, and the fact that a few United States corporations dominated the field. How did Canadian policy-makers respond to computer technology? The computer industry can be defined as having a character completely independent of national context, yet paradoxically, its effect cannot be divorced from specific national conditions.[1] This ambivalence has been a vital one for Canadian governments; it strikes at the heart of the Canadian technological experience. When confronted by these new technologies, governments first acted ambiguously. Should they do something, or nothing at all? When policy-makers eventually responded, they did so by applying well-tried historical models about the proper relationship between new technologies and government policy. Hence the focus on issues of foreign ownership and investment rather than technological leadership, and how to promote innovations in niche markets, for example. By the late 1960s, the federal policy towards the computer industry was a mixture of ambiguity, lack of coordination between policy and government purchasing, and a lack of political commitment to change the status quo. Governments seemed to be frozen between the urge to meddle and do something on the one hand, and the impulse to stand back and let it all happen on the other.

THE ORIGINS AND DEVELOPMENT OF COMPUTER TECHNOLOGY IN THE UNITED STATES

Three elements combined to establish American dominance in the computer industry. First, a successful combination of public and private initiative established the development of computers as a national priority. Second, huge public investments and private capital were amassed to fuel both innovation and market development. Third, both the state and the market took advantage of relatively rapid technological change in the electronics industry. The cooperation between public and private interests pushed the us computer industry to global leadership in the field, and established a single, global marketplace.

As with innovations in communications and power in the nineteenth century, computer technology moved along established south-to-north channels, although the current was not always one way. The affiliation was natural, offering the Canadian economy the blessings and perils conferred by American links. The evolution of the American industry towards globalization bore profound consequences for Canadian policy. Initially eager to diffuse the blessings of technological modernity, Canadian policy-makers eventually succumbed to concern about the implications for statehood, control, and national identity. By the time the alarm was raised, the moment was lost.

In order to understand developments in the computer industry, it will be worthwhile to outline its growth in the North American marketplace. The electronic digital computer represented an enormous technological leap, specifically by reductions in calculation time. A combination of better system architecture and improved components has meant a fivefold order of magnitude (10^5) improvement in computer-processing speeds since the 1950s. But the principal architecture was put in place by the mid-1960s, and entrenched by the explosion in commercial computer use in the early 1960s. The sustained support of the us Defense Department, combined with the steep decline in the cost of computer capacity, positioned the technology to enter the market. The alliance of national security and new technology guaranteed the computer a fast track.[2]

Cooperation between public agency and private firm in the United States consolidated the global leadership of the American computer industry after 1955. Government-industry partnership also generated a large unified market for its software and propelled penetration of foreign markets with American computer shipments. This coincided with an expansion of American foreign investment in the mid-1960s. The combination of the two alerted the governments of the industrialized West to the perils of the American challenge.

There are some remarkable similarities between the emergent Canadian computer industry and the more established communications sector in the 1960s. Both sectors were dominated by institutional leviathans – International Business Machines and Bell Canada – who continued to control development of the industry well into the 1970s, when technology would undo both, or at least compel them to reinvent themselves. The pattern of technological development in both sectors was dominated by the singular fact that two giants controlled the field, at the height of their monopoly control. Smaller entrepreneurial firms worked the margins of the industry. In the computer services sector, time-sharing computer networks and artificial intelligence applications stood as examples of special product niches that the big companies were unwilling to develop.

In telecommunications, the trend in computerization reduced the costs of operations. Statistics available for the United States suggest a 9 per cent per year decline in incremental transmission costs between 1952 and 1965.[3] Flamm argues that the pace of advance slowed after the early 1960s, achieving minimal decreases from 1965 to 1970. This suggests a slower rate of technological progress in telecommunications than in the computer sector.

An explanation for the differential rates of progress in the telecommunications and computer sectors might lie in the regulatory structure and the nature of the unchallenged monopoly. But the fundamental technological difference between communications and computers might best explain the differences in rates of decrease. Computers race against time; communications battle distance. Given Canada's transcontinental expanse, it was natural that improvements in transmission were given pride of place over speed.[4]

In Canada, both telegraph and telephone networks were harnessed. Out of telegraph technology came TELEX and TWX. Telegraphy had the virtue of converting information to coded electromagnetic pulses more cheaply than the telephone system could. The demand for digital transmission encouraged common carriers to expand their switching capacity.

The alliance of communications and computer sectors created an embryonic information sector, including communications equipment, computers and electronic products and software, as well as telecommunications and computer services. The explosion of long distance telephone service joined with advances in the commercial computer sector to create a powerful phenomenon. In 1965, for example, IBM engaged in discussions with Bell Canada to convince them that the recently introduced System/360 would solve some of the serious bottlenecks that were occurring in the Bell system. The reason that such a discussion could be held at all was the convergence of technologies. An internal IBM memorandum compared the system architectures of the System/360 and the No.1 ESS electronic switching systems and found that they were largely compatible when it came to data-processing capabilities. The company recognized that convergence would be an increasingly integral component to IBM's business.[5]

By the 1960s the American computer industry had established virtually worldwide dominance in the field, with IBM as the major player. Comparisons between the computer industry and the more established communications sector illustrate the truly remarkable growth of computer technology. In the next section, we will examine the consequences of that growth for Canada, and how the Canadian state struggled to come to terms with an increasingly important economic and technological phenomenon.

CANADIAN RESPONSES TO THE COMPUTER, 1948–1965

In Canada, as elsewhere, the rapidly looming technology summoned policy-makers to an exceptional historical moment. How would the new technology be harnessed? The computer had defined the moment; the state now had to respond. The industry was now demonstrating its economic power to continentalize, and it did not take much to imagine the technology's political implications. These same implications were giving European political and technocratic elites sleepless nights. Policy-makers in European capitals viewed American technological leadership as a threat to their own efforts at mounting an effective challenge to pervasive American economic and cultural influence. Perhaps just as important, European governments also recognized the implications technological lag had for their ability to promote indigenous technological research and products. It was in the 1960s that this vital dimension became an important political question. On this terrain, a peculiarly Canadian response was fashioned – a response that would set a definitive pattern for years to come.

What characterized Canadian thinking about this new technology? The answer is complex, since it involves the interplay of several distinct currents. First, Canada's tradition of resource export dependency, combined with its import substitution industrialization, resulted in a technological dependence that resonated throughout the twentieth century. The advent of computer technology confirmed the fact yet again. Second, governments and computer industry players were induced into a game of catch-up as a result of the rapid diffusion of computer technology. What is more, by the mid-1960s questions about Canada's technological sovereignty and fears of a technological backwardness that would hamper economic development were coming to the fore. Canadian political discourse on the subject would be increasingly dominated by the belief that the technology was a crucial component in economic growth and international competitiveness.

Canada's long experience with regulation shaped the state's reception of computer technologies. With telecommunications, conflicts over control of what was considered common property and the exercise of monopoly power ended up in an array of regulatory responses early in the century. State supervision was established.[6] Later in the century, computer firms found themselves more successful in Canada in defending themselves against direct regulatory oversight. The computer industry (principally IBM) had staked a preponderant claim, and successfully defended the field against significant state intervention.

The Canadian computer industry grew up very much in the shadow of its American superiors. In the late 1940s a few Canadian businessmen came together to found Computing Devices of Canada (CDC). The pattern of government support exhibited in the United States was replicated: CDC survived largely on research and development contracts from the Canadian navy and the Canadian airforce. The company found it difficult to produce major innovations; instead, it obtained the Canadian franchise for certain US designs. In 1952, CDC purchased the rights to the CRC 102A, developed by the California-based Computer Research Corporation. The CRC 102A was primarily geared towards engineering and science-related activities. That same year, Avro Aircraft of Malton, Ontario, bought one, followed by the RCAF for use at their base in Cold Lake, Alberta. Canadian defence needs drove the early industry, both in telecommunications innovation and the development of computers.

Outside the military-industrial complex, the University of Toronto was at the leading edge of the technology with the development of a relay-type computer in 1948, and by 1950 it began to teach computer technology and applications to its undergraduates. However, it took IBM's introduction of the IBM 650 in 1956 to herald the definitive arrival of the computer in the commercial life of the country. By 1960 the government had all but conceded that "it seems fair to say that practically all the initiative in computer development in this country thus far has come from outside Canada, particularly from the United States." To be sure, the Defence Research Board had been conducting research into a prototype computer, and Ferranti-Packard was striving to develop an airline reservation system.[7]

The transformation of government and business to electronic systems was also inhibited by high purchase prices of the large systems. Clerical salaries were lower than in the United States, so computerization was not as appealing. Why have a room full of expansive vacuum tubes when you could have a room full of clerks? As well, Canadians thought themselves to be "innately more conservative than the Americans and less willing to adopt new techniques." The state of the technology favoured the larger establishments. They accounted for approximately half of the electronic data processing (EDP) installations, whereas the small establishments accounted for one-fifth.

The unfolding of computer technology stirred national policy-makers to respond. As early as 1955 the federal government took its first tentative steps towards advances in electronics. A committee was formed by the Treasury Board in March 1955 to "study developments in Electronic Computing machines and their application in the Public Service of the Government of Canada." New functions and new mechanisms led to a transformation of relationships between the state and the public, which would have an impact on later decisions about the nature of the technology. The committee's philosophy regarding computers was a cautious one, but one that was equally mindful of the potential benefits of the computer, especially in matters of administrative efficiency. "The mind of man is often slow to grasp all its [sic] potentials of a new tool in its hands," the assistant dominion statistician wrote, "[f]irst, he must work with this tool and then gradually explore its uses."[8]

The Cold War and the launching of Sputnik in 1957 seemed to make the use of high technology a patriotic duty. The launch prompted some to re-examine priorities in education and research and development. As a Canadian professor, Basil Myers, warned in 1960:

The concern of the public [about technological lag] has been motivated almost entirely by fear. It is just as well that this is so because, as is common knowledge, this fear has been justified ... We now find ourselves launched in what may truly be called the Space Age. We find that our position as a nation in science in technology is not unique. We are particularly concerned with *survival*, and it is this aspect, more than any other, which has brought education into the limelight in very recent years.[9]

"There is no use in kidding ourselves about it," he added, "[w]e are bound to conclude that the *Russians have a clear edge* in the quality and purpose of their over-all educational system."

By 1958, a survey of private and public sector organizations noted a significant increase in the use of computing devices, and states that they were beginning to be used more and more in commercial applications.[10] By 1960, the status of electronic data processing in Canada drew the interest of the Department of Labour and the federal government's Advisory Committee on Technological Changes, which was set up in 1957. Concerned with technological changes in key industries, the Department of Labour conducted an important survey of EDP in Canada. The subsequent report provides an important indication of the condition and course of the Canadian computer industry.

The study reported that at the beginning of 1960 there were only eighty-nine electronic digital computers in operation. The geographic concentration of the technology was pronounced: most computers were located in Ontario and Quebec, mainly in Toronto and Montreal. The smaller-scale computers were used almost exclusively for scientific work, while three-quarters of the large computer capacity was used for commercial data-processing.

The report further noted that electronic data-processing in Canada had created over a thousand full-time jobs in Canada that had not existed before. EDP organizations were especially top-heavy: over half were administrators, project planners and programmers. This was contrary to the perception of "the svelte blonde seated at the control of the 'giant electronic brain.'"[11] Only at the junior level was the female complement more noticeable.

A second survey of technological change in Canadian offices was done in 1962, and it noted that the electronic data processing was the "most recent but potentially most far reaching of all the changes that are taking place in the office environment." The subsequent study noted the prodigious development of the sector: a trebling of the personnel employed in full-time EDP, growing shortages, and an exploding computer population. If anything, the report underestimated the diffusion of electronic data processing in Canada. There was hardly any reference to two Canadian companies most concerned with computing equipment – Computing Devices of Canada and Ferranti Limited. The report also overlooked the fact that by

1961, over nine hundred students had attended computer courses at the University of Toronto. As a result, the report was coolly received by the newly formed Canadian Information Processing Society.[12]

The Canadian market could not hope to match its American counterpart in realizing the promise of electronic data-processing. The Canadian market in 1962 was conservatively estimated to be only 3 per cent of the American, with a more feeble rate of growth as well. However, the expanding importance of the industry was beginning to be appreciated. One observer warned that electronic data processing had "arrived at a point in history when the progress of our times could not advance at full speed without it."[13]

Faced with the expanding burden of paperwork and communications, administrators were increasingly relying on electronic aids. Developments in administration made it necessary to "think through the cloud of fear, prejudice and hysteria that surrounds the progressive steps of automation." The implications for economic development would be critical. Those hesitant to accept its importance were directed to look to the immense efforts of the Communist world in automation to find illustration of its importance.[14]

If there was proof of American predominance in leading-edge technologies, it was IBM's introduction of the System/360 in 1964 which reinforced the idea of Canada's technological drag. Writing in *Fortune* magazine in September 1966, Tom Wise enthused: "The new System/360 was intended to obsolete [sic] virtually all other existing computers ... It was roughly as though General Motors had decided to scrap its existing makes and models and offer in their place one new line of cars, covering the entire spectrum of demand, with a radically redesigned engine and an exotic fuel."[15]

Although the idea of a standardized, compatible family of computers was a few years old, IBM was the first to introduce it, and reaped the rewards. Tables 12 and 13 aptly illustrate the remarkable extent to which the company had grown.

The problem of American penetration in a key sector elicited a variety of responses across the globe. The solution popular in Europe of creating replica IBMs foundered precisely because competing with IBM in its established markets proved to be an ultimately futile strategy. Even state barriers to entry in the form of tariffs and restrictions to the market could not deter American predominance.

EDP had become as important as the telephone to the smooth operation of modern business. Computers could resolve the mounting processing and control problems increasingly identified as expensive obstacles to productivity. The difficulty was that the mountains of paperwork had to be converted to a machine-readable format. By 1963 the *Financial Post* could report that the computer had assumed an "essential place in the spectrum of business equipment" and would within brief compass eclipse all the other tools modern business had used to date. Most of the thirty universities in the country had installed them for teaching and research purposes; in business, the value of computer plant was put at over $100 million. Computer service centres were handling more and more workload for Canadian businesses.[16]

The prospect of connecting computers with communications began to find expression in the business discourse of the day. Innovation and demand compelled both the computer and telecommunications industries to consider new service offerings. The more specialized demands of customers required the exploitation of the telecommunications voice network for data transmission. AT&T president Frederick Kappel predicted in 1960 that data traffic on telephone networks would exceed that of voice transmissions by the end of the decade. If Canadians could not expect that to occur in their country, they could be assured that the amount of data transmission would rise exponentially.[17]

By the mid-1960s, the diffusion of computers drew increasing notice. The appearance of a moderately priced medium "solid state" computer pitched to commercially-based data processing unquestionably influenced the expansion. Canadian business was increasingly concerned with the break-neck pace of technological change. The computer was heralded as the most significant development since the introduction of electric power. In an article published in the *Harvard Business Review* but widely reprinted in Canada, James R. Bright warned of the opportunity and threat that lay coiled within the computer core:

Thousands of businesses are going to rise or fall on the ability of their managers to respond effectively. How can management understand this environment, and how can they meet it wisely? ... There is no certain way to success and security in this technological ferment, but leaders of all types – industrial, military, political and social – must become more skilful in dealing with technological change. I believe that the first requirement for the businessman is a keen sensitivity, awareness, and receptivity to technological changes as a major environmental force which he can employ, and to which he must respond ... How long will it be before, say, American Airlines is challenged by "American Communications Inc?" It seems to be that dealing with technological innovation is becoming a more serious, a more frequent problem for all of society as well as for business management in particular.[18]

The possibilities were dramatically underscored by IBM's national and international performance, which culminated in 1963 with the introduction of the System/360. This third generation of computers was an outcome of what the company called Solid Logical Technology (SLT) and replaced the earlier second generation models that had been introduced in 1958. System/360 would target government and business users alike, peddling such applications as welfare accounting, tax processing and information storage, and law enforcement. The company chose to manufacture its Canadian requirements at its Don Mills, Ontario, plant. The announcement would unleash a protracted battle for position in the Canadian market, and stack the odds in favour of IBM.[19]

The extension of IBM's operations and the growing diffusion of electronic data processing attracted the interest of the Department of Industry. The department's Electrical and Electronics Branch was the vantage point for the government's consideration of the Canadian computer industry. The branch's view of the industry was blurred, and even vaguely apprehensive. IBM had brought with it the distinct advantages of advanced technology, availability, and reasonable price. Some officials suggested that acquiescing in American predominance of the industry would result in throwing away Canada's potential to offer its own contribution to the computer market. In many respects import substitution had worked well, but its side-effects were producing some worry about Canada's long-term ability to produce in a crucial sector.

The growing prominence of the computer industry was certified by the numbers: the world market size in 1960 was less than $1 billion; in 1967 the industry grew to $6.6 billion. Predictions for 1975 were for a global market of $39 billion. Those numbers were encouraging European governments to act. In Britain, a network of subsidies and university research sought to position British industry to compete with IBM through its national champion, International Computers Limited (ICL).[20] In Canada, the rapid development of the value of the computer industry was averaging a rate of 25 per cent per year. Yet no indigenous Canadian industry existed; the technological activities were limited to selling imported equipment by American subsidiaries. IBM had a manufacturing plant in Toronto which handled mainly spillover production for the Canadian market.

Canada's presence in the field had been aided by the establishment of the University of Waterloo in 1957. The university's strong link to the computer industry illuminated both the national promise and the pitfalls of entering an occupied field. The creation of the university was partially predicated upon the necessity of closer industry/university ties. To them, industry- and business-oriented research had not been a priority for Canadian universities. Both IBM and federal government support underlined the practice by funding pure science. The change of attitude was reflected in Waterloo's first president, Ira G. Needles. Drawn from the chairmanship of the board of B.F. Goodrich Canada, Needles's mission was straightforward: to strike an alliance between the university and industry. The unabashed goal was to assimilate the technologist and engineer into a university environment. That commitment to collaboration with industry took on major proportions, and included undergraduate cooperative programs in which students were given opportunities to both study and work.

Perhaps the most striking development at the university was its establishment as a centre of computer expertise. The establishment of a computer centre and the devotion of the Department of Mathematics resources to the practical problems of computer development fitted nicely with the mission-oriented nature of the university. Under the leadership of J. Wes Graham, and Mathematics chairman Ralph Stanton, computer science became a central feature of the university curriculum. Its importance led to the establishment of a Faculty of Mathematics in 1967 which emphasised applied research.[21]

The technological environment created at Waterloo was the direct consequence of the coalition between business and learning. IBM provided the cooperation and hardware at a substantial discount. Professors and administrators shared the goal of developing specialization in a domain of ascending importance. The result was a number of important achievements in educational software. The WATFOR compiler was designed for an IBM 7040 computer, meant to reduce the processing time of a student's FORTRAN program from over a minute to less than a second. WATFOR was soon used in applications all over the world. The National Research Council provided a significant proportion of the funding. The province also recognized its importance by awarding massive financial assistance to capital construction and computer equipment installation. Waterloo was soon advertised abroad as a typical example of Ontario's technological modernity.[22]

Waterloo's rise to prominence is instructive as much for its successes as the limits it suggests about Canada's technological position. The university's ability to establish itself as a centre for computer expertise depended in large part on the presence of a North American technical and capital pool represented by IBM. With a combination of entrepreneurship and research, Graham, Stanton and others involved in computing sciences managed to exploit an undeveloped dimension of a market dominated by one large American company.

Commenting on Waterloo's reliance on IBM equipment, Graham offered the following explanation:

The situation as I see it is as follows. The IBM company has the vast majority of the business in the computing field ... In terms of universities, where we are installing computers, one of the questions that always comes up is: if you install a piece of equipment for educational purposes should you or should you not install the piece of equipment which is in most popular demand?.. It is quite a vicious circle which could have, perhaps, unfortunate consequences ... [but] They happen to do many things right and they happen to prosper and they have a lot of things going for them.

In some ways, the Waterloo experience illustrated the parameters of the possible. On the one hand, the university provided a warm climate for Canadian technological research in application software; on the other, the industry-university partnership of the 1960s was of necessity continental in nature. The former gave policy-makers plenty of cause to celebrate; the latter gave them some reason to be concerned.

The nagging fact of American dominance remained. IBM Canada imported its data processing systems from its parent company. Few Canadian components were used in systems assembled in Canada. The Canadian branch had responsibility for the manufacture and engineering of the 6400 accounting/computing machine, but it was not considered to be a "general purpose digital computer." What worried Department of Industry officials was the increasing disparity between the company's exports and imports. Despite over $31 million in exports, the company imported over $34 million. The department predicted that the imbalance would reach $60 million industry wide in 1966 and a possible $100 million in 1967.[23]

The expanding importance of the sector in national productivity and its absolute dependence upon foreign technology and manufacture attracted the closer attention of the Department of Industry. The department predicted that if trends in computer trade continued, the existing imbalance of $85 million in 1967 would grow to $450 million by 1975.

Reduction or even elimination of that deficit would be desirable economically and of significant national advantage. Since IBM commanded over 70 per cent of the market, working with the computer industry was virtually synonymous with working with IBM. In 1966, Canadian businesses had paid nearly $140 million for IBM's goods and services, and in the first three months of 1967 alone, they paid IBM $40 million. Indeed, the company had outfitted the federal government with one of the most advanced and largest data processing operations in use anywhere to streamline and centralize administration.[24]

In 1967, the minister of industry, C.M. Drury, along with his principal advisers S.S. Reisman and D.B. Mundy, met with IBM Canada's president, Jack Brent, to convey the federal government's concern over the situation, and to suggest remedies. Discussions between the minister and IBM executives paralleled Electrical and Electronics Branch deliberations with IBM officials, but they failed to secure any meaningful concessions from IBM. The department complained of IBM's serious lack of cooperation. The company would not provide quantitative information about its Canadian activities, preferring to deal in generalities. The only way IBM would move fast was if the government was interested in buying.

The department's efforts to stimulate competition and market entry even extended to encouraging Britain's ICL to consider the Canadian market for expansion. ICL declined, preferring to restrict its Canadian involvement to supporting Ferranti in any sales that company might be moved to take up. ICL was too concerned with combating IBM on its home pitch, and could not entertain mounting a serious challenge to IBM dominance overseas.[25] Tables 14, 15, and 16 demonstrate the fact that computer imports were far outstripping computer exports by 1966, a situation which became progressively worse in ensuing years. The effect on the office machinery industry in Canada was noticeable.

The hope of encouraging a European manufacturer to establish itself in Canada had evaporated with ICL's rejection, and the Department of Industry's options had been dramatically reduced in the process. Departmental objectives were twofold, and related to the manufacturing and distribution of computers in Canada. First, it sought to establish that industry targets in Canadian manufacturing value-added were commensurate with domestic consumption by 1973. Second, the department insisted that the industry develop a research and development capability in Canada equal to the ratio of global consumption to global R&D. That meant the federal government expected by 1973 that the $150 million consumption in the Canadian computer market should be matched by a "value added" to the Canadian economy of the equivalent amount. The alternative: that Canadian technological input would be limited to a "special engineering facility supporting sales."[26]

In 1967, only $150,000 of IBM's research and development met the government's R&D guidelines. In effect, the Canadian computer market was supporting foreign R&D activity, which was estimated at between $12 million and $25 million. The Department of Industry argued that if that R&D were repatriated, it would support the deployment of between three hundred and six hundred scientists and engineers. But with moral suasion and little else at their immediate command, the department could do little. The alternatives would be to either allow present trends to continue, or establish joint ventures with other foreign firms and information utilities with Bell-Northern and CN/CP in a consortium, with the government as leader.

With computer use growing exponentially, and IBM responsible for about three-quarters of the equipment, the Department of Industry had good cause to target the colossus. For IBM, the status quo had produced big profits and little or no research and development commitments beyond a pledge to manufacture some items locally. The persistence of this situation would mean that Canadian computer firms would lose the opportunity to bring Canadian labour and scientific resources to bear on a quickly developing and important technology. Inaction would also exacerbate the serious disparity in balance of payments in the sector. IBM Canada would buy from the American parent at one-third the cost. Canadian companies, on the other hand, were required to pay full duty on every component that came across the border. The tariff structure clearly favoured both National Revenue and IBM. For its part, the Department of Industry hoped for a quid pro quo from the company.[27]

Given the circumscribed nature of state action, Industry officials were limited to suggesting that awareness and self-interest were crucial. IBM management had to be made aware of national problems and priorities and somehow link them to a rationalization of their activities which would benefit both the country and the parent company.[28]

Canadian research and development policy tried to stimulate subsidiaries that were able to develop special products for the domestic and global markets. If the Department of Industry could get the branch plants to specialize rather than to act as a production unit, Canadian niches would be opened up. The bulk of research would be absorbed by the US market, giving Canada access to the most advanced technology.[29] Getting national and IBM policies to coincide was a laudable goal, but the tools to make it so were limited. Federal government action was restricted by consideration of government procurement policies and manipulation of the tariff structure to provide incentives for Canadian subsidiaries to establish data processing activity. Most of all, it would require voluntary action by the company. The policy would not address the problems of mediocre management in the field and foreign control, but it might provide a point of departure. The basis was thought to be that all programs and actions of the department should cause men in the industry to think and make sound business decisions, and in so doing, achieve government aims.

The computer was originally a product of the office and business machine industry. Because of the single company dominance of the sector and the secrecy provisions of the Statistical Act, data on the computer industry are difficult to segregate from the entire industry. Nonetheless, the Department of Industry estimated that computers were responsible for approximately half of the industry's revenues, or $200 million in 1966.[30] The fact remained that most of the Canadian market for computers was supplied by imports. Canadian content was approximately 40 per cent, but that content was at the technological lower end of the scale – manufacturing and the like. Industry officials were not optimistic that the situation would change unaided by government intervention. Only the intercession of the state could stem the American technological tide.

A measure of the ambiguity of the federal government towards the situation in the computer industry was reflected in the gulf between attitudes and action. By end of decade, the Department of Industry had gone from curiosity about the industry to demands for greater computer research, development, and manufacturing. Policy-makers were convinced that Canada was not participating in a key technology and was not likely to without state intervention. That situation would produce two outcomes. First, inaction would result in an inability to hold indigenous scientific talent. Second, Canada would remain technology-dependent. Yet the federal government took great pains to provide assurances that the industry would not be disturbed. The minister, C.M. Drury, was anxious to dispel any association that increased state intervention might have with French and German dirigisme, signalling that if state intervention would come, a light touch would be applied.

Government procurement might have provided just such a light touch. Regrettably, lack of policy coordination neutralzed it. IBM had moved as early as the late 1950s to establish a predominant position as the principal provider of computing power to the state. Purchasing was done at the departmental level; if one convinced the department, the sale was made. The Royal Canadian Air Force, the Dominion Bureau of Statistics, the National Research Council, and other departments turned to IBM for computing. The company also ensured that it maintained strong links to the Canadian Data Processing Service Bureau, (CDPSB) the central government clearing house for computing. IBM's Ottawa office had turned the capital into an IBM town. As a result the government's national policy agenda and individual departments' purchasing programs were often in conflict.[31]

A dramatic example of the possibilities procurement offered was a project undertaken for the federal government's Agricultural Rehabilitation and Development Agency (ARDA). The agency desired to computerize the physical resources of the country. Since the Canada Land Inventory Project was a massive undertaking, ARDA approached IBM to execute the project. By late 1967 it was evident that IBM had overshot its budget on the project. Reasoning that IBM would gladly be rid of the project, and that the federal government wanted to expedite the project, three IBM employees formed a consulting firm to complete the ARDA project, using IBM computers. In March 1968, Systems Dimensions Limited was formed. The new company would become a major participant in the Canadian computer services industry.

The opportunities generated by government procurement were eventually recognized by entrepreneurs. The creation of Systems Dimensions and Computel attested to the drawing power of federal government business. The new companies convinced the government that they could better manage most of the government's data processing, and could do it cheaper as well. As a result, the CDPSB was disbanded.

The federal government's experience in procurement is instructive. The lack of a coherent policy meant that departments of government were often working at cross-purposes. Concerns about national technological potential at the Department of Industry were neutralized by the keen wish of other departments to get the industry standard, and at a good price. This led to a situation where an IBM salesman could sell the federal government on the need for a large data processing system, then go into private business and subsequently convince the government to contract most of their work out to the private sector.[32]

Despite the lack of an explicit policy, government procurement did nourish an indigenous computer services industry. The industry was created with state help, but with a state virtually oblivious to what it was creating. Used consciously, procurement could have been an instrument to meet some of the state's objectives in the field. Preoccupied with the giant, the state failed to give proper consideration to the more modest but effective devices at its command.

Canadian policy-makers had also turned their attention to time-sharing computer centres. Originally proposed in the United States in 1959, the concept developed principally in the academic community, and was typically supported by US government research funds. Time-sharing services allowed remote terminals to access a central computer's power. As early as 1960, the promise of a super-system of time-shared computers was firing the scientific imagination. Writing in 1967, Thomas Pyke marvelled at its potential benefits:

The creation of a super-system of time shared computers could, if properly designed and used, link together enormous amounts of knowledge, with the work of many thousands of people in many fields. If the system could be made to assist in the organisation of knowledge so that it is easily accessed, it could be the communication link necessary to keep up with the technological world which is enveloping man. It could have untold amounts of duplicated effort and bring together disciplines that may otherwise never have found common ground.[33]

Network connections could offer multiple connection, permit several terminals to use transmission lines alternately, and enable the multiplexing of data transmission.

Time-sharing streamlined accessibility of computing power by eliminating the problem of distance. The new developments also provided the attractive possibility of using a computer in "real time."[34] The set-up seemed perfect for smaller users who could not afford their own large systems but required the capacity of a large computer to perform periodic tasks. The airline reservation system SABRE (Semi-Automated Business Research Environment), introduced in 1963, was the most illustrious example of time-sharing technology. The spread of digital transmission capacity over the public telephone network improved its efficiency.

By 1968, the computer service industry in North America was offering more and more business-oriented services. The configuration of the time-sharing phenomenon provoked earnest discussion of an information utility. Fisher, McKie, and Mancke recall that as early as 1966, Western Union was advertising a complex "designed to provide information, communications and processing services in much the same way as other utilities supply gas or electricity."[35]

By the late 1960s, the federal state could point to the emergent computer communications field as an area in which public policy could exert its influence. If it was too late for the state to have a major impact on the configuration of the computer equipment industry, the government could enter on the ground floor with the marriage of computers and communications. Here, regulatory control over communications was already firmly established. Past sins of omission in technological development could be atoned for by a more vigorous direction of this new field. And the state could participate in the guidance of the new technology for national goals in economic development and national integration. The attributes assigned to telecommunications – its economic importance, its relevance to national unity and sovereignty, and its important implications for Canada's international position – were easily enlisted to describe the intrinsic value of the marriage of computers and communications. Here the state could most effectively promulgate its goals with an added technological thrust. The integration of the state's historic aims in telecommunications – notably universal service and coverage of distance – were grafted onto new technologies which promised speed. Paradoxically, the demand for technological sophistication spurred by the same technologies would later lead to increased demands for value-added services and a more flexible structure that would undermine the traditional pillars of communications policy.[36]

CONCLUSION

By the end of the 1960s, some key trends in computer technology had already become fixed. An American company had established global dominance in the field. There was little that Canadian policy-makers could or would do to counteract this. The benefits of this process were manifested in the rapid diffusion of the technology across the Canadian market.

The state was not as slow to grasp the relevance of the technology as its actions might suggest. Certainly by the mid 1960s, questions of statehood and national control began to surface, particularly in its dealings with IBM. The borrowed nature of Canada's technological advance was most apparent here.

Why then did policy-makers defer to the status quo? First, the Canadian market was relatively well served, so there was little consumer pressure for the state to intervene. Second, implementation of a vigorous program of state intervention in the sector would have taken a commitment might have had disastrous effects. The political will for remedial legislation barely existed. Third, the state was scarcely aware of the possibilities of fostering a stronger presence in the field through the instruments of procurement and the tariff. Had a government procurement policy existed, the state might have been better able to direct the growth of the industry through its own buying. Had the tariff been lowered on computer components from the United States, the Canadian computer industry might have been in a better position to compete with its continental rivals. Fourth, there was a sense that the die had been cast in the computer industry; IBM had created a technological fait accompli. Proof lay not only in the numbers, but also in the fact that the state itself bought large numbers of the company's products, and relied on IBM for Electronic Data Processing advice. As a result, a marked ambivalence characterized Canadian policy towards the computer industry.

If the door had closed to the computer industry, the emerging field of computer communications must have seemed wide open. This new area offered the state an opportunity to ride a technological wave and to leave a stronger imprint of the national interest in a pivotal technological development. Exploiting technological opportunities would in part depend upon the regulatory structure as well as how the political machinery of state perceived the dynamics of change in telecommunications. To better understand this challenge for the state, the next chapter will take a look at how the Canadian government responded to technological transformation in communications in the 1960s.

4 Revolution and Reaction: Telecommunications Policy, 1960–1969

In his study of the dynamics of technological change, Thomas Hughes uses the notion of "salients" and "reverse salients" to explain technological transformation. He defines a salient and its reverse as a "pronounced projection or bulge in an advancing front; a reverse salient, an oxymoronic concept, refers to a part of a front that lags behind."[1] As the last chapter demonstrated, the federal government's ambivalence over the issue of technological leadership minimized its role in the development of a key technological industry. In the Canada of the 1960s, state responses to changes in telecommunications were the reverse salient in the system, which delayed the effective management of telecommunications policy issues.

Once a peripheral issue, communications policy moved onto the central agenda of government in the 1960s. Two major technological problems pushed the state to develop a policy: the establishment of a second microwave network, and pressures to launch a communications satellite. By 1968, telecommunications issues ultimately led to the establishment of the Department of Communications. Departmentalization of the issue demonstrated the importance of the matter, and the government's desire at least to appear to be in charge. The field attracted a group of ambitious and talented policy-makers dedicated to generating policies and programs where there had been none before. Forming a coherent national telecommunications policy would later require negotiation with both industry and the provinces.

By the 1960s, technological transformation and growth in service demand brought the industry to the verge of some striking changes that challenged established patterns of pricing and fuelled the modernization of plant. Improvements in switching, transmission, and terminal equipment emphasized the possibilities of providing cheaper services and creating a range of new ones. Consumers were demanding more, and technological change was offering the possibility of providing more. This combination meant that telephone companies across the continent were finding now the demands for technologically sophisticated services difficult to keep up with.[2]

An additional element in prodding the state into coming up with a coherent policy in telecommunications was the general impulse to modernize state administration. The administrative machinery of the state had come under increasing scrutiny and was found wanting. Together, technological change and administrative lag caused such a serious strain on policy that new institutions better suited to the task had to be created.

Here we will explore the idea of technological progress in the Canadian context. The federal response to developments in domestic telecommunications in the 1950s and 1960s and to international arrangements over satellite technology are the main concerns of the following sections. The chapter will also discuss the pressures leading to the formation of the Department of Communications in 1968, and particularly the federal government's attempts to defend its interests in the domestic sphere over the question of satellites.

THE CANADIAN IDEA OF TECHNOLOGICAL PROGRESS

There has been little attempt in North American history in general and Canadian history in particular to explore the nature of technological progress in the post-Second World War period. Nonetheless, clarifying the issues surrounding technological change is an important task. Technological progress as a concept is as important as its concrete manifestations. Both the measurable dimensions of the technological revolution and the role it played in the public imagination were important conditions in the development of a policy framework for telecommunications. The aura of revolutionary change created by communications technology produced public spectacle and induced public interest. This theatre of revolutionary technology found a public well disposed to its devices. The message responded to a desire for novelty and a need to know that technology could be controlled. Communications technology provided the assuring paradox of experiencing something new that was already familiar.[3]

This was the case with the technologies that moved into the public eye in the 1960s: expanded opportunities and advances in telecommunications, broadcasting, and computer technology were wondrous, but not alien. Speculation drifted in the public mind from the possibilities of global community to the prospect of a world where housewives and drop-outs, the homeless and lonely millionaires, could share their experiences. Such was the stuff of the theatre of technology. Another major tenet of the technological canon was the notion of living in an accelerated age. The acceleration could be dramatic. Anthropologist Margaret Mead could write by the mid-1960s that "the gulf separating 1965 from 1943 is as deep as the gulf that separated the men who became the builders of cities from Stone Age Men." The rush of technological advance was driving the world at "apocalyptic speed" whose effects were at best uncertain.[4]

In Canada, the worry was not only acceleration but also the fact that Canadians were watching events from a kind of Stone Age of their own. George Grant combined a critique of North American culture and capitalism with an examination of the role of technology and concluded that technology had become the "dominating morality." He argued that Canadian society had surrendered to a technological imperative that was out of our control. But even Grant himself accepted prevailing opinion that technology was all-encompassing and definitive.[5]

The idea of rapid change produced some giant leaps of faith. The prophets of the electronic revolution cast themselves in the role of "secular theologians" in the phrase of James Carey.[6] Technique would be able to do what past transformations were unable to achieve – remove barriers to social harmony. Electronics would have more potential for social change than the railways of the nineteenth century, and electricity of the twentieth.

The claims for the technological revolution were as numerous as the enthusiasts who celebrated them in automation, nuclear power, space travel, telecommunications and computers. All of these technical changes would bring the transfiguration of economic, social, and political life that were associated with the British Industrial Revolution. This technological revolution would usher in the post-industrial society. For the most part, these passionate predictions were based on speculation rooted in technocratic rationality and perceptions of past transformations as revolutionary.

The business press of the period was also full of metaphors of technological acceleration and change. Both political and business discourse about technology stressed both the threat of obsolescence and the gift of opportunity. Writing in the Harvard Business Review in 1963, James R. Bright warned of the consequences:

Unquestionably, our era of dynamic business change is based on technological progress. In this mercurial environment, traditional products, materials, skills, and production facilities are made obsolete in a few years and in some cases a few months ... Thousands of business are going to rise or fall on the ability of their managers to respond effectively ... These new demands do not present themselves nicely at the doors of obsolete businesses, nor do potential disruptive innovations announce their birth with trumpets. There is no certain way to success and security in this technological ferment.[7]

More importantly, these changes promised economies of effort and certification of progress which had usually secured a kind reception in North American history.

For Canada, the hope of a technological tomorrow was tempered with hesitation about its implications for national survival. If Canada failed to master its communications problems, one official wrote, "Canada will have lost her essential raison d'être and will disappear as meaningful entity." Communications might also have virtuous political consequences: conditions were believed right for the creation of a "hotline democracy" to "make things happen" and give form to what was called "people power." What is more, if properly harnessed, communications would give youth a chance to be heard, mute the separatist impulse, and create the "world's first truly bilingual, multi-cultural society." It was increasingly evident to policymakers that the problem of communications was as vital to the survival of the country as transportation was the century before. The danger was that telecommunications technology might outstrip the country's ability to assimilate it. "The difference between what may and what can happen," Eric Kierans wrote in 1969, "will be decided by political will."[8]

There were three principal federal objectives in telecommunications: to ensure that telecommunications did not become a barrier to favoured economic activities; to promote overall political and economic objectives; and to guarantee the use of the telecommunications infrastructure in the promotion of state social and cultural interests. The decision-making structures in place in 1960 were increasingly unequal to the task of ensuring those historic objectives.

American historian Kenneth Bickers has argued that technological innovations may provide an important corridor between incremental politics and the politics of transformation. He suggests that "new technologies raise the question of how the innovations should be treated, and by whom. A struggle may emerge over the issue of whether the new technology should be subsumed under the inherited industrial categories or should be viewed as the basis of an incipient new industry."[9] Questions of technological direction in Canada began as a whisper in the late 1950s, and became a shout by the late 1960s. Canadian state policy in telecommunications would begin as incrementalist, but end up as transformative. Advocates of change in regulatory design for the telecommunications industry had mobilized the authority of the state to the cause of reform. It would become evident, however, that this effort to expand the state's administrative capacities would be incomplete.

THE FEDERAL MANAGEMENT OF COMMUNICATIONS ISSUES, 1957–1965

By the mid-1950s there were two focal points for the formulation of telecommunications policy: the Privy Council Office, and the Telecommunications and Electronics Branch of the Department of Transport (DOT). The Privy Council Committee on Telecommunication Policy was formed in 1957–8 and began life dealing with policy problems referred to it by either cabinet or the Department of Transport. The committee was originally created to consider one specific telecommunications problem: the potential civilian use of military communications systems such as the Mid-Canada Line and SAGE. The committee was increasingly called upon to consider a range of telecommunications matters and was eventually given permanent status under the chairmanship of R.B. Bryce. The committee advised the cabinet on such matters as the CN/CP proposal to install a second Trans-Canada communications link.

Despite its increasing prestige, secretary E.F. Gaskell lamented the committee's limited sphere in drafting telecommunications policy. Planning and policy power still lay with the Telecommunications and Electronics Branch of the Department of Transport, effectively preventing a comprehensive outlook. The branch had pursued a policy of benign neglect, to the detriment of overall planning. The result was a failure to define the interests of the state on a variety of questions, from competition to ownership of communications facilities.[10]

The policy vacuum had tangible consequences. The common carriers claimed that it was difficult for them to "plan for the future" or to "work in close co-operation with the governments such as ENTO [Emergency National Telecommunications Organization] planning and northern coverage." Their real concern was to forestall any moves towards extending government ownership. National ownership, Bell argued, "can only have detrimental effects upon the industry itself, because the opportunity to provide the communications services which it is so well fitted to supply is reduced." The industry further lamented the state's presence as a competitor through CNT and COTC, which came at the expense of the Canadian taxpayer.[11]

The Telecommunications and Electronics Branch generated most of the worry, since it retained general responsibility for telecommunications matters, including licensing, emergency planning, research, and links with the International Telecommunications Union. Its mission of guaranteeing orderly growth in facilities was implemented by soliciting voluntary compliance, establishing franchise areas where none existed, and administering the Radio Act.

The condition of Canadian telecommunications policy drew the attention of the Royal Commission on Government Organization of 1963, headed by J. Grant Glassco. The commission's views reflected emerging disagreement over the proper scope of government participation in the telecommunications infrastructure. The commission concluded three things. First, the power to make telecommunications policy was based on the Radio Act. National policy could not be created on such a narrow basis. The commissioners cited the cabinet decision to establish a second microwave link as proof that telecommunications policy had to be put on a new footing. Second, the gross duplication evident in the Telecommunications and Electronics Branch could be alleviated by streamlining. The existing policy of separating radio frequency and common carrier regulation prevented telecommunications policy from being considered as a whole. Third, Canada's international interests were being compromised because of understaffing and lack of expertise. The commissioners cited Canada's meagre participation at international conferences, particularly the 1959 ITU Radio Conference, as proof that the challenges of protecting the national interest were not being met. Continued confusion here would expose the country to greater problems in the future.[12]

Planning for new technologies without proper authority made matters worse. The federal government confronted technological change with a mixture of inadequate legislation and semi-independent, inadequately coordinated operating groups, sparse technical knowledge, and scanty resources. That sometimes resulted in bad economic decisions, strained relations between the state and the carriers, and bad coordination of federal-provincial efforts. Having a comprehensive policy – one that would be modest, but strategic – might be the solution.

The commission's worries were not shared by F.G. Nixon, director of the Telecommunications and Electronics Branch. From his point of view, the branch laboured hard persuading common carriers to provide maximum national coverage with minimum duplication. The branch supervised the provision of military, air traffic, marine, and meteorological services. Much of this was done with no explicit mandate. If the existing statutes were being used in a "questionable manner," it was getting the job done. The branch used its licensing power to ensure, for example, that its policy of maintaining three service areas in the North was respected. But dealing with telecommunications matters of "daily increasing complexity" would take a lot more than the power to grant licenses. The situation was an embarrassment to the commission, but not to Nixon.

The commission's insistence on cleaning house at the Department of Transport was understandably met with resentment. Nixon contested the need for a higher authority to formulate telecommunications. He argued that the department had made itself clear on every issue. It simply was not practical to enunciate policy, he argued, since technologies were changing so fast. Nixon eventually conceded the desirability of having some direction from higher levels for telecommunications policy, but only after a great deal of convincing by royal commission officials. In spite of the twin pressures of rationalizing government and technological change, the department's top telecommunications official had to be persuaded forcefully that there was a need for policy co-ordination.[13]

The problem of technological change alone provided the rationale for its importance for government:

The entire telecommunications – electronics art is characterized by rapidity and complexity of growth, diversity of application and impact upon every phase of human endeavour. In order for Government to keep pace with their dynamic evolution, constant reviews and a periodic management audits of the type conducted by the Royal Commission on Government Organization will be essential.[14]

The growing doubts about the effectiveness of the federal government's telecommunications policy were confirmed when a major policy problem surfaced. In 1961 the federal government had to deal with the question of the construction of a second transcontinental microwave system by Canadian National and Canadian Pacific Telecommunications.

Competition between CN/CP and the Trans-Canada Telephone System over long-haul transmission had intensified in 1954 when the

Telephone Association of Canada (TAC) established a Halifax-to-Victoria high-capacity microwave telecommunications system. In the same period CPT and CNT jointly established a similar high-capacity system between Montreal and St John's. In mid-1961 CN/CP argued that general economic expansion and increasing demand for telecommunications services meant that a second microwave network was necessary. Failure to approve the new system, they argued, could eliminate them from competition in the public communications field. CN/CP hoped to obtain assurances that the government would "give all the business which could reasonably be given to this system if it is competitive in cost with similar services being provided by the TAC." The railways were determined to capture the COTC's business, which by 1964 would include thirty-six high-quality voice circuits to connect the Trans-atlantic cable to the COTC's Pacific cable. The Department of Transport was hesitant, uncertain of its ability to provide the necessary government business to the system. It also feared it could not guarantee healthy competition. The COTC lobbied for the system, but suggested the government acquire part ownership. In any event, they were making policy without serious direction.

The question of a second microwave system was debated within the Privy Council Committee throughout 1961, with final cabinet approval coming in 1962. The government defended the project by invoking military necessity, but commercial considerations were just as important. Actual construction of the network began in the summer of 1962. To satisfy defence and commercial requirements, the proposed route was well north of transcontinental facilities. The system was completed in May 1964, at a cost of $42 million. Its promoters heralded it as part of a "dynamic new era in telecommunications, highlighted by the evolution of machines and miracle techniques" which had been unfolding since the late 1950s. The microwave network consisted of 136 towers between Montreal and Vancouver, designed to augment the burgeoning traffic stimulated by the use of telex and private wire services. CP also planned high-speed data transmission links, with switching centres established on the new microwave network. By 1967 CN/CP had opened the Broadband Exchange Service, a data transmission facility which took advantage of CN/CP's "microwave super-highway" completed in the mid-1960s.[15]

The microwave network cast doubt upon the state's capacities. Policy-makers found themselves barely able to deal with the econo-technical complexities of installing a second transcontinental microwave system, which only represented an incremental change in technology.

The state's experience with the second microwave network converted some senior policy-makers to the necessity of managing technological change in telecommunications. Microwaves alone, however, were not enough to budge the state to systematize its approach to telecommunications. If policy-makers still needed persuading, the arrival of the satellite would leave no room for doubt. The multifaceted nature of satellite technology would compel the state to develop a framework for telecommunications planning.

THE STATE AND THE SATELLITE: PRESSURES FOR A NATIONAL TELECOMMUNICATIONS POLICY

The arrival of satellite communications technology gave physical form to the hopes for technological advance. The story of satellite development in this country has been well covered.[16] It is more difficult to discern the effects these technological transformations had on state power. At the very least, the satellite prompted the state to clarify its general interests in telecommunications.

The Canadian government's experience with satellite technology was rooted in military cooperation with the United States. The Defence Research Telecommunications Establishment (DRTE) had already participated in scientific satellite ventures with the United States in the late 1950s. By 1962, Canada had designed, built, and launched the Alouette I satellite to take measurements of the ionosphere. Alouette II followed three years later, in November 1965.[17]

More important for commercial communications were international efforts to establish a single global communications system. Demand for intercontinental communications had outgrown existing facilities; this assured the commercial viability of any future satellite system. Increased demand also necessitated a response from the federal government that it was ill-prepared to make.[18]

The cabinet had agreed to financial participation in an American scheme, but with two conditions: Canada would have a say in the supervision of the system within the context of an international agency, and technical and economic conditions should be satisfactory. By early 1964, interested Western European governments met with US and Canadian officials in Rome to advance the satellite idea. The general objective was to create a global system "in the spirit of relevant United Nations resolutions." The Rome agreements concerned the activities of the newly formed United States Communications Satellite Corporation. The satellite it proposed to launch would be ready by 1965 and provide 240 voice circuits, depending upon the number of ground stations. Another satellite with full global coverage was planned for 1967. The estimated cost was over $200 million, and was to be paid by the United Kingdom, the United States, Canada, Japan, and Australia.[19]

The final organization of the scheme was cause for concern in both Canada and Europe, who desired the establishment of a permanent world organization to administer the satellite. Both recognized the superiority of American technology over their indigenous efforts and were anxious to participate in a satellite construction that would provide business for their industries. This would bolster European and Canadian chances to reach full technological sovereignty. On the other hand, the Americans were anxious to retain control, advocating a management scheme that gave voting rights in direct proportion to investment. That left Canada with 5 per cent, the Europeans with 30 to 35 per cent, and the Americans with most of the rest. Both Europeans and Canadians were unprepared to accept this, favouring instead some system of weighted voting which would limit the dominance of the United States.

The American plan for launching satellites for both civilian and military telecommunications was also a cause of concern. First, the proposal would mean that Canadian satellite needs would have to compete with US military and civilian use. That was especially troublesome because there would only be one North American receiving station. Second, the proposal was only partially designed for commercial requirements, and Canada's domestic market might not be well served. Third, the scheme might supersede Canadian plans for using transportable ground stations to relay the satellite signal to give Canadian forces a potential military advantage. Lastly, it would upset Canada's plans to develop industrial capability for spacecraft construction.[20]

The question of corporate management prompted more general concerns about American control. The American undersecretary of state, Eugene Rostow, argued that the satellite carried no ideological goal. Rather, it would symbolize "the truth that space belongs to all men, that an international undertaking that permits the free flow of communications."[21] Based on its experience with international communications issues in the postwar period, Canadian policy-makers had good reason to be skeptical of Rostow's claims.

The desire for the technology overpowered their anxieties about foreign control. The Europeans were the most anxious to participate in the design and development of the satellite system. To that end, they favoured an allocation of production on the basis of financial contribution, together with the purchase by the consortium of US patents. Canada favoured a more open competition for production, especially since its share in the corporation amounted to only 5 per cent. The Canadians were more interested in satellite capacity.

Uneasiness about Canada's ability to answer these challenges prompted the establishment of another committee in early 1964. The Interdepartmental Telecommunications Committee was struck to deal with the mounting evidence that a more systematic way of dealing with communications had to be found. It considered the matter serious enough to recommend that cabinet establish a royal commission on telecommunications activity. At the very least, chairman Gilles Sicotte argued, there should be a thorough examination of activity in the common carrier field. It was thought that the absence of regulation was "detrimental to the interests of carriers and users alike." Canada had allowed the telecommunications industry virtually to regulate itself. This was no longer acceptable, for the public interest and the national interest were not well served by the status quo. The issue of foreign control of telecommunications, though not an immediate danger, was cited as a prime example of what might occur if the federal government did not proceed with reorganization.[22]

The committee recommended that the government create consultative machinery that would not tread on provincial jurisdiction. As an alternative, it proposed that the federal government regulate long distance, and leave local telephony to the provinces. A more radical solution would have put all telecommunications under federal control. Pressure for a greater focus on telecommunications mounted throughout 1965 and 1966. The subject had been discussed by the cabinet in the early 1960s, but that discussion dealt only with the international participation in American satellite proposals. A clearer picture of Canadian domestic interests emerged by the mid-1960s.

In July 1966, Bell Canada applied to the minister of transport for permission to build an earth receiving station for satellite signals. In October, southern Ontario broadcaster Niagara Television sought permission from the Board of Broadcast Governors to launch a satellite system to distribute television programs for a third television network. The TCTS and CN/CP Telecommunications were also proposing to launch a system for telecommunications. The federal government proceeded to bring the analytical resources of the government to bear on a problem over which the private sector had already clearly given much thought.[23]

It was becoming clear that satellites would simply not provide additional network capacity. With the launching of the Early Bird Intelsat I satellite in March 1965 as the first commercial satellite by a US-led international consortium, the Americans signalled their desire to capitalize on recent advances in the field. President Johnson's aggressive message on the possibilities of satellite communications in the summer of 1967 did not go unnoticed in Ottawa. How would Canadian voices be heard over the din? The technology might consolidate American commercial predominance. The Oval Office advised the Canadian government in July 1967 that American intentions included the erecting of a single global commercial systems of at least three satellites. The operation of the satellite would be on a for-profit basis. That would mean direct competition with older forms such as cable and terrestrial radio systems, and subject to central direction.[24]

Meanwhile, the results of studies conducted throughout 1967 and 1968 concluded that 50 per cent of the space segment and 95 per cent of the ground segment could be designed and produced in Canada. The problem was the considerably higher Canadian cost. With several major questions left unresolved, the government released the White Paper on Satellite Communications in March of 1968. The intention to proceed with a domestic satellite communications system was affirmed in principle.

The possibilities of the satellite were not lost on the Quebec government. The province conducted negotiations with France throughout 1967 and 1968 over a possible joint satellite venture to service la Francophonie. The Quebec initiative was well received by the French government, especially since their space program was in need of backers. Quebec's participation would also support the French thesis that regional satellite communications networks were as viable as the global system promoted by Intelsat. The project to procure satellite capability – dubbed Memini (je me souviens – I remember) – would also help Quebec's strategy to pursue negotiations with international bodies in order to gain leverage closer to home.[25]

Quebec's persistent desire to participate in international satellite projects left External Affairs in a quandary. On the one hand, it was anxious to preserve the federal state's exclusive jurisdiction over international telecommunications; on the other, Ottawa wished to accommodate the province's concerns. "It could lead to the notion," a member of the Canadian Intelsat delegation wrote to Allan Gotlieb, "that whenever Quebec claims jurisdiction we jump in and give them a place on Canadian Delegations."[26] The problems were rendered more difficult by the precarious state of France-Canada relations.

Quebec's reasons for seeking French partnership were compelling. Agreements were already signed between Quebec City and Paris on manpower training. The province hoped that it would promote expertise in satellite construction, and amplify Quebec's place in a global French communications network. A satellite link with France

could strengthen its Quebec's cultural position. The risks of not participating were also clear: "le risque sera grand d'une sorte
d'asphyxie culturelle et la position même du français s'en trouvera
un peu plus affaibli," wrote *Le Devoir* in 1967. Negotiations with
France should proceed, with the hope that other countries of la Francophonie would be able to participate, according to their financial
ability to do so. Premier Daniel Johnson had visited Paris in the
spring of 1967, and the establishment of closer ties to France had
generated much enthusiasm. It had also generated heightened tensions in the already strained relations between Ottawa and Quebec
City over questions of both carriage and content.[27]

By the time of its launch date expected in 1972, Radio-Québec envisioned a programming schedule that would reach out to the world
via satellite. The "Symphonie" project would assist the province in its
drive to modernization and without any protracted negotiations or
compromises with Ottawa. Symphonie's sphere of operation would
cover the eastern half of Quebec and part of the coastal United States.
One commentator noted with evident satisfaction that "[l]'Ontario est
donc completement exclus du rayon d'action du satellite."

The satellite question encouraged the province to define its interest
in telecommunications and broadcasting. Provincial officials envisioned its organ, Radio-Québec as standing at the centre of an electronic communications system – the "Quebec branch of a world
knowledge bank." The plans of Quebec officials clearly raised questions about jurisdiction and the utilization of new technologies.
Quebec was beginning to argue that communications and culture
were inseparable in the electronic age, and the latter was an extension
of a provincial responsibility. This directly challenged the national
conception of its own mandate, and the clear line that Ottawa drew
between the cultural and the technological.

A major thrust in Quebec's desire to ally itself with a German-
French satellite venture being negotiated in Ottawa was the creation
of an indigenous technological workforce within the province.
French-Canadian engineers and scientists could harness their training in the service of extending the technological frontier. Nationalists
had long complained that Quebec was underfunded in terms of
research. There were considerable pressures from the province's universities and related sectors to assure the province's participation in
the "forward march of science and technology," and Ottawa was well
aware of the dilemma. It also fuelled the drive to establish provincially controlled satellite communications.

Canadian External Affairs officials in Paris were watching these
developments closely. In late 1967 the embassy in Paris reported that

the French had become doubtful of Quebec's capability to participate in a joint satellite venture. Apart from RCA in Montreal, Quebec industry had insufficient experience for such an undertaking. Contrary to their earlier impressions, the embassy reported that Quebec would not be allowed to join the Symphonie project. It was clear that the Quebec government had seriously underestimated the financial and technological details of such an undertaking. For example, the province had been dismayed to learn that France expected the province to put up over $40 million as the price of participation.[28]

Meanwhile, Ottawa prepared to defend its claim to exclusive jurisdiction over satellites by concluding an agreement with West Germany, France's partner in the satellite project. John Chapman of the Defence Research Telecommunications Establishment argued that it would clearly be "to the advantage of the federal government to establish itself, by formal agreement, as the sole authority in Canada with whom the German Government can deal in space matters, research development, production and operation." Chapman wagered that, given a choice between "causing mischief" in Canada and continuing good relations with its German partner, France would opt for the latter.[29]

The growing friction between Ottawa and Quebec City over the satellite issue was indicative of the problem of technological transformation: what was to be the form and the control over domestic communications. It also presaged the development of a full-scale confrontation over jurisdiction a few years later. Jacques Gauthier, adviser to the Quebec government in communications issues in the late 1960s, aptly summed up the prevailing sentiment in Quebec. He argued that in the nineteenth century imperialist countries divided up territory; in the late twentieth century the risk was that the most prosperous nations would do the same with space. Quebec was simply affirming its jurisdictional competence in an area where its survival was at stake.

If the opportunity for participation in the Franco-German satellite was lost, it did provide the Quebec government with greater ties to Paris and a stronger position at the negotiating table in Ottawa. Although keeping the French connection open for future consultations, Premier J.J. Bertrand underscored the importance of allying with Canada and presenting a "front uni afin de ne pas être soumis au jeu des grandes puissances dans l'occupation pacifique de l'espace par les ondes et les satellites." The conflict, however, was to express itself more forcefully over two central issues: that of cultural sovereignty, and who would own and build the satellite.[30]

TELECOMMUNICATIONS POLICY COMES OF AGE: THE CREATION OF THE DEPARTMENT OF COMMUNICATIONS

The realization that the administrative and policy-making functions of the federal state were inadequate to the task of managing increasingly important political and economic questions arising out of satellite technology led directly to the creation of the Department of Communications (DOC). Announcing the new department in 1968, Prime Minister Trudeau also stressed the need to organize government activity in telecommunications policy. As its first minister, Eric Kierans told the Commons that communications was new but profoundly important to the future of the country. If Confederation had been bound by mile upon mile of steel rail, satellite communications would perform the same task for Canada's next century. Technology would provide what politicians could not: a bilingual nation with a northern vision. Such were the hopes for technology in Canada.[31]

The DOC was fashioned out of a patchwork of branches from other departments. The Telecommunications and Electronics Branch was transferred from the Department of Transport. Internal telecommunications policy and the management of the Radio Act, the Telegraphs Act, the COTC Act and the Railway Act were all put under the DOC umbrella. Also transferred were the duties and functions of the Science Secretariat in relation to the satellite communications system, as well as the Defence Research Telecommunications Establishment of the Defence Research Board. It amounted to over seven hundred person-years of resource transfer and control over a budget $7 to $8 million.[32]

Creators of the new department aspired to redraw telecommunications policy. The Post Office was to be combined with the new department, but was removed after fears were expressed that this signalled the beginning of a European style PTT (post, telegraph, telephone) organization, which was highly centralized and state-controlled. Department objectives were explicitly formulated to confront growing problems of technological change. Computers, personal privacy, the relationship between communications and transportation: all were issues that would command the attention of the new department. The means was a national communications policy that would break the "sorry pattern" of old thinking. Public and private sectors would work together by producing the technological expertise to meet the demands of the Canadian communications system, and ensure national cultural security.

One of the more important dimensions of the new department's duties was promoting research. The DOC was empowered to oversee that communications research was "adequate to the national need" by sponsoring research in communications and conduct reviews about the requirements of both private and public sectors. Up to 1968, the research and development component had been given only haphazard policy attention, despite the best efforts of the defence researchers. Worry about falling behind in R&D was powerful enough to stimulate great concern in cabinet in 1970. Kierans informed the cabinet "there is a danger that Canada will lose control of events and miss the potential benefits of an intelligent exploitation of [communications] techniques." There was a need for government investigation of subsystems and large-scale systems alike. Industrial capability in Canada in areas such as applied digital technology was limited to a few companies. Kierans recommended that the DOC be more directly involved in the formulation of scientific research in the government and its projects be assigned priority.

The new department was also expected to foster the "orderly development" of national telecommunications. In addition, it was given the task of management of the radio spectrum, taking that function over from the Department of Transport. For the first time there would be some systematic thought about the development of "principles relating to regulation of communications, and of mechanisms for so doing." That would be the most challenging task for the department. Telecommunication legislation was scattered across several statutes that had been passed over a long period of time. Policy had to step in where fortuitous cooperation had operated in the past. To meet the challenges of modernization, the DOC would devote much of its energies to achieving a "total regulation" of telecommunications falling under its jurisdiction. Federal control had to be effectively extended to all works under its jurisdiction, in order to apply technology to national goals. This also meant tracking the public need for communications and the ability of Canadian industry to serve those needs. Economic and commercial viability and the harmonization of the regulatory environment were paramount.[33]

Radio spectrum management and regulation did not have the cachet that the shaping of information technology did. Kierans's executive assistant, Richard Gwyn, argued that the name of the department should have been changed to the "Department of Information Technology" because it provided a more accurate description of its activities. Gwyn meant the discussion to signal his desire to see the department at the frontier of technology – a highly desirable place for the department to be, in a domain of escalating importance.

Gwyn and others wished to utilize the department to coordinate, promote, and recommend national policies, something which had never been tested. The act that created the department provided for the creation of a national policy, but it was seen as strictly window dressing by its authors, its purpose being to let other departments and the public know the general intentions of the government. But the very success of the department would ultimately depend upon how well it would guide developments in telecommunications and computer/ communications.[34]

From its inception, the Department of Communications aspired to become more than a simply technologically oriented department which the Department of Transport had been in telecommunications issues. The new department struggled to develop a comprehensive communications policy commensurate with the importance popularly attached to the field. The legislative mandate was clear: the department should promote the orderly development of telecommunications in Canada. That effectively meant that the grand objectives of Canadian telecommunications policy would be the provision of cheap, reliable, and accessible telecommunications. Within that context, the DOC had to tackle several pressing issues. All of them were connected in one manner or another to the larger goal of preserving and promoting order in telecommunications, and ensuring that national objectives in the field were met.

Realization of the department's objectives was seriously hampered by the lack of a clear direction beyond its initial optimistic pronouncements. Since the DOC was starting from zero, even "raw data for decision-making" was lacking. As Richard Gwyn wrote to the deputy minister, "[a] great deal of information is tucked away in people's heads or desks and is trundled out, verbally and randomly, when particular crises arise."[35] Organization and systematization were critical if the department had any pretensions to exerting influence over the revolution.

The need for such a plan was manifest to the neophyte department for several reasons. First, the expansion of communications systems in Canada led to the conviction that communications carriers could only be preoccupied with short-term objectives of running the system. As a result, their activity often overlooked the larger national context. The DOC concluded that the greatest influence on industry could be exerted if the department concentrated on influencing the company's day-to-day short-run plans and programs, but with national goals firmly in mind. For that to happen, the operations side of the DOC aspired to become "extensive in nature," with the power to "really know what [was] going on."

Second, the heralded arrival of a "wired city" connecting everybody by telephone, computer, and cable influenced policy-makers. The wired city would feature a large central machine, and a corresponding network of wide-band cables, in which the spectrum would be sectioned between broadcasting and computer services. Subscriber needs would determine the type of service received. The new reality would place immense demands on public policy. Policies were needed on common ownership of carrier facilities, foreign ownership of wired information networks, and the technical and organizational aspects of the distribution networks.

Changing the regulatory structure for telecommunications was perhaps the most important task facing the DOC. Pressures for change in the regulatory environment from groups increasingly dissatisfied with the lack of restraint on privately-owned public utilities, as well as the strain induced by the introduction of new services and technologies, were pushing the state to define the public interest, and to act. The point was driven home to cabinet in 1969 by CN/CP's purchase of a computer information company, statements by CATV owners that they were ready to become common carriers, and the arrival of the Canadian satellite system. Those events highlighted the need for a definitive telecommunications policy.[36]

In leaving technological advances for policy responses, we leave the world of superlatives for that of modest qualifiers. The DOC was hampered by a wavering political will to promulgate a national telecommunications strategy. The act gave the minister of communications important powers to recommend, plan, and enact national communications policies. But, as we have seen, it was not a mandate to pursue radical initiatives. The most optimistic officials within the department, however, still stressed the promise of how the DOC could occupy a major position within various sectors of the communications field. For those officials, the success or failure of Canadian communications policy rested upon the creation a powerful legal and regulatory directorate.

One thing was clear: a comprehensive study of telecommunications was necessary. Its goal would be to recommend elements of a national policy and the legislative means to that policy. The ultimate objective would be to overcome the vacuum of federal planning that had been filled by the industry itself. As the department's deputy minister, Allan Gotlieb, noted, the industry "ha[s] to a large extent exercised functions which in all other western nations are considered to be a prime responsibility of government." The piecemeal and outdated approach to regulation was in danger of locking the state out of a "knowledge transmission system" which would link Canadians with each other at low cost. In leaving things as they were, the state could be eliminated from participation in the most momentous changes of the twentieth century.

Inevitably, the target for any major federal telecommunications policy initiative would be Bell Canada. Many of its activities were regulated by the Canadian Transport Commission, and the agency's preponderant influence over the Trans-Canada Telephone System made it a natural focal point for potential federal action. By the later 1960s, the company began to recognize that the social and political dislocations of the decade might actually have an effect on them. "While Bell appears to do business in much the same old way at very much the same old stand," one Bell official commented, "the environment around us has been changing, by public consent, to one that our fathers would have found unbelievable and our grandfathers unacceptable."[38] As was the case with the regulatory component, the company had to be vigilant against unpalatable changes that might interfere with smooth company operations. By 1968, some Bell officials were calling for a re-examination of traditional concepts and corporate aims in order to suit the times. To that end, Bell had to be seen as making a positive contribution to the public interest, and not simply reacting negatively to proposals for change to telecommunications policy.

As early as 1968, Bell officials began to be wary of federal plans for the telecommunications sector. Department of Transport officials reassured the company that their views would carry much weight, and promised full disclosure of any proposed changes. Bell did not hesitate to tell E.R. Bushfield and Gordon Haase of the department, however, that the company would fight any attempt to "cream-skim" the more lucrative parts of its business, or degrade the integrity of the network. The task before the company lay in convincing policy-makers that excessive liberalization would operate against the public interest, "which would be best served by Bell retaining control over the network in both the technical and economic senses."[39]

The company nonetheless recognized a sea-change in the federal government's attitude towards a range of telecommunications issues. It made its intentions clear: it wanted to see a "faster rate of introduction of new technology" by loosening restrictions on competition. In the matter of interconnection, the government was determined to modify the legislation. The company had to be prepared to defend its role by demonstrating what it had already achieved by way of the provision of "high quality, plentiful, economical, convenient, and reliable telecommunications services." The company argued that total end-to-end control had been a major factor in the excellent performance of the telecommunications industry in Canada. Any potential changes should be implemented slowly, and with deliberate care.

Despite the bravado about facing technological and political change, Bell could not hide a certain trepidation about the possibility of telecommunications reform. The establishment of the Department of Communications and the special task force was interpreted as an unnecessary interference with the status quo. As the Bell's general supervisor of regulatory matters discerned, however, the company would not be "too adversely affected" by federal attempts to strengthen federal authority in the telecommunications field. "Canadian telecommunication legislation is in a rather sorry state," he commented, "and there is little logic to the present arbitrary division of jurisdiction between the Government and the Provinces in telecommunication matters." Some systematization was necessary and perhaps even desirable. In the meantime, Bell would have to move beyond simple restatements of the excellence of the Canadian telecommunications industry and defend the industry's interests.[40]

Bell's initial strategy would be resistance. Any attempts to restrict Bell's role in the total communications picture and promote "wasteful" competition would be met with opposition. The company was also concerned with demonstrating how the public interest was served by telecommunications in everything from business to defence to national unity. Attempts at tampering with Bell's successful formula might have disastrous results in innovation, service, and cost.

The centrepiece of the company's strategy was naturally the preservation of their position of "natural" monopoly. The purpose here was to show that the public interest would best be served by having only one supplier of communications services per territory. New carriers, most notably cable television companies, should be limited to the pick-up and re-transmission of broadcast program material. Developments in the telecommunications field had made it technically possible to support a diversity of telecommunications services into the home, but what was technically possible might also be economically disastrous. Demonopolization through liberalized policies over interconnection, for example, would translate into costly duplication and cream-skimming, and jeopardize the Canadian telecommunication system.[41]

Those planning company strategy over the creation of a national telecommunications policy were surely comforted by the federal government's limited sphere of action in telecommunications. Any comprehensive telecommunications policy would have to be negotiated with the provinces, and, as J.A. Harvey, Bell's vice-president for

regulatory affairs noted, would perhaps require "prior recommendation by a Royal Commission in order to impart authoritative support
to the general concept." Bell would have ample time to exert its
influence and anticipate possible changes well in advance.[42]

The company was not taking things for granted, however. Bell's
private and monopolistic nature had caused public suspicion and distrust in the past, as had its apparently close relationship with the
regulator. The company's position on a range of issues, from federal
income tax payments to interconnection, had created a surprisingly
broad range of adversaries. Bell had to show that it was "in the public
and national interest for Canada to continue to have this big, powerful
communications company as the leader of the industry." Only titans
could hope to remain in a world dominated by large organizations.
By implication, Bell argued that excessive state intervention could not
possibly be in the public interest, since it would weaken the giant.

THE CONFLICT OVER A DOMESTIC COMMUNICATIONS SATELLITE SYSTEM

The federal state's experience with the Franco-Québécois satellite
question strengthened its determination that technological changes
had to be infused by national purpose. A Canadian satellite had to
be put up. Dr John Chapman, author of the report that had forcefully
recommended the launching of a nationally-based satellite, justified
it primarily in terms of national sovereignty:

Authority over domestic communications is vital to any state and full control
is desirable, since experience has shown that Canada cannot rely entirely on
the us or any other country for essential supplies and services during times
of emergency ... Furthermore, Canada needs to exert sovereignty over the
Arctic archipelago, and other northern territories, which entails adequate
communications.[43]

Chapman was supported by the Science Council of Canada, which
also stressed control over domestic communications. Canada should
not wait to stake its claim: the United States and South America
would be competing for the same "parking space" in orbit. The consequences of not proceeding might well be loss of control over the
Canadian telecommunications system. The fragmentation of Canadian sovereignty by independent provincial action had to be forestalled. As Allan Gotlieb remarked, "Our major objective at the
moment is to be up there first by a couple of years and then it would
be to everyone's advantage to deal with the Canadian satellite."[44]

By late 1967, the question was no longer if, but how. The hot questions were over financing and control. Niagara Television and Power Corporation proposed a $75.5 million satellite for domestic communications. The common carriers countered by proposing an $80 million satellite that would be used with existing radio-relay and other terrestrial facilities. The Canadian embassy in Washington warned that "wealthy and powerful groups" in the United States were moving quickly to secure as much of the field as they could. The course that the government would eventually choose was therefore of "no small importance." In March 1968 the government announced its intention to establish a domestic satellite communications system, to be owned and operated by a corporation with mixed ownership.[45]

The policy had several objectives: the provision of nationwide television service in English and French, and extension of service to unserved areas; the supplementing of data service and telephone service in areas already served; and stimulation of Canada's industrial capability to control the design and production of the system as much as possible. Linking Vancouver Island and Newfoundland, Pelee Island and Ellesmere Island, would be a "wonderful and vital modern development," a benediction of science for which the country would be grateful. Getting in the procession for space was therefore important.[46]

The ownership structure of Telesat Canada, the crown corporation that managed the domestic communications satellite system has been exhaustively detailed elsewhere. For the purposes of this book, it will be relevant to recall a few of the salient details as it affected the general canvas of communications policy in the late 1960s.

Questions of who should own and control public utilities have usually generated controversy. In the case of telecommunications, federal officials argued that the telecommunications industry was "socially and economically vital to the country." It was therefore perilous to allow a carrier-owned satellite communication system, since it would reinforce Bell's strong position in the domestic marketplace. Instead, federal policy-makers proposed a crown corporation which would have the advantage of introducing new services and technology without having to protect considerable investments in existing physical plant.

Crown ownership would also have the advantage of ensuring some kind of competition between carrier and non-carrier suppliers in the building of the satellite, ground stations, and terminals. It could also ensure maximum Canadian participation in satellite construction, and strengthen Canadian industrial capability in the international marketplace. The government understood that the

telephone companies were prepared to match the government purchase of equity of $30 million. Another $30 million would be offered
for public issue in one or two years.

The common carriers were concerned that the satellite would form
the nucleus of a third communications network in competition with
their own. They had not had full use of their microwave systems and
were anxious to put off a commercial satellite venture until they
recovered costs of the older system. Carrier fears were reflected in
their dealings with the state; at one time they threatened to pull out
of the project. In the face of a firm stand by the federal government,
the common carriers ultimately participated.[47] TCTS was forced to
cooperate with the federal government which would not back down
on the question of state participation.

The public-private participation in the corporation demonstrated
the state's desire to distribute the risk of an undertaking that required
huge capital outlays. The original proposal included a tripartite
scheme where government and the common carriers would divide
60 per cent of the shares equally, with 40 per cent being sold to the
public. The common carriers countered with a proposal that
excluded the public shares provision, at least in the initial issue of
Telesat Canada shares. Although provisions for the public share
made it into the final legislation, the state and the carriers held virtually all of the shares. The corporation was a political holding company – an "expression of a political economic bargain struck to
generate greater certainty and control in an industry besieged by
technological changes and conflicting policy purposes."[48] The advantage of being in the vanguard of new technologies had to be balanced
with preserving the viability of the existing facilities.

The cabinet gave approval in February of 1969 for a six-channel
satellite and a system of ground station at an estimated cost of
$66.2 million. In May, Kierans and his department were authorized
to enter into discussions with RCA and Northern Electric with a view
to securing the maximum amount of Canadian business for the
satellite. Agreement in cabinet over Telesat ended there.[49]

The reaction to the cabinet's decision was overwhelmingly favourable to the idea of a national satellite; the structure of ownership was
the problem. If, as the government contended, the satellite would
have profound political and sociological implications, and the communications industry was the central nervous system of Canadian
society, then public ownership was clearly indicated. Allan Gotlieb
defended the Telesat decision before the Commons Standing Committee on Broadcasting, Film and Assistance to the Arts. The intent,
Gotlieb explained, was to "create a society, a corporation where the

government would have the kind of special responsibility that comes with being an owner, but it would share that with private enterprise in the form of the carriers and it would share that with the public."[50]

Eric Kierans, the minister of the DOC, justified the purchase of some American technology for the Canadian satellite by reminding the Commons that the technology being used in the satellite had "come really to an end in terms of the type of satellite that was being built." Canadian participation would come in the planned communications technology satellite and Canadian industry would be able to participate as a central player in that initiative.[51]

Soon after the establishment of Telesat in 1969, the corporation surprised many in the cabinet by accepting the unsolicited proposal of an American company – Hughes Aircraft of California – over the Canadian bids to design the satellite. The reasons were simple: Hughes built a better satellite, and it had agreed to have a certain level of Canadian content. The Hughes design also offered increased potential revenue, greater operational flexibility, and made more rigorous contractual demands on the company. Kierans vigorously supported the Hughes bid, and took pains to remind his cabinet colleagues that the management of RCA had a spotty track record, especially in terms of cost overruns and poor business decisions. RCA's promise that it would make major changes to company organization specifically for the satellite project did not convince Kierans. The company also threatened 250 immediate and one thousand additional lay-offs if it did not get the bid. The company reminded the press that the Hughes bid had approximately one-tenth Canadian content. This directly contradicted the white paper's conclusion that "the control, specification, design and construction can and must be retained in Canadian hands."[52]

The issue of who should build the ground segment for the new satellite complicated matters. It also touched a number of issues, not least its policy of maintaining a vigorous Canadian presence in a new field, considered of major importance for the future for the Canadian communications industry. Canadian capabilities in the sector had been enhanced by contracts from Intelsat.[53] Exports of telephone and microwave relay radio systems were good and the industry was also hoping to gain significant ground with satellite earth station systems.

Industry hopes were frustrated by the drift of government policy over Telesat. The president of the Electronic Industries Association, Léon Balcer, registered his strong dissatisfaction over Telesat's procurement strategy for earth station requirements. Letters of interest were sent to American, Japanese, and European contractors. This was particularly galling to Balcer, since the contractors' home governments only considered their domestic contractors for such undertakings. A "penetration bid" could be used to enter the Canadian market and establish a permanent presence.[54] For its part, the Department of Communications was caught in a similar dilemma as in the Telesat bid: to seek the support of Canadian industry and the strengthening of Canadian technological capability, or to secure the most cost-effective and rapid service.

Kierans's support of the American bid unleashed a storm of protest within the cabinet. Many of his colleagues, supported by an increasingly hostile press over the government's satellite plans, argued that since satellite technology was new, creating a technological base was paramount. Kierans countered that Northern Electric was neither aggressive nor "initiative minded" enough to win the project, and SPAR Corporation would have taken too long. RCA had overextended itself in the field of colour television sets and its ability to branch out from the production of phonograph records to six or twelve channel satellites was in doubt. Canada could save money with the Hughes bid, and spend it to better advantage in the field of communications.[55]

A three-day fight resulted. Quebec ministers, led by Bryce Mackasey and Jean Marchand, emphatically argued for the RCA bid for reasons of political necessity and technological sovereignty. The government had spent over $20 million since 1963, with the hope of fostering research and development. Although RCA had a substantial piece of Canada's share of the communications technology satellite project directed by NASA, the likelihood was that RCA would pull out of the space business if it lost the Telesat bid. Commercial viability and exploiting maximum industrial potential seemed to be mutually exclusive. Hughes had produced the Syncom and Intelsat series as well as other communications satellites. Kierans warned that if the government withheld approval of the Hughes proposal and accepted the RCA one, the launch date of the satellite would be delayed. Further, the RCA proposal would mean that the federal government would have to provide all the capital, or at least half, and dramatically reduce Telesat's ability to generate revenue. The cabinet was overwhelmingly in favour of the Quebec position, but the prime minister and the PCO secretary, Michael Pitfield, were convinced that since Kierans was taking the risk, he should also make the decision. Kierans was so committed to his plan that he would have resigned had he lost the battle.

Similar circumstances plagued the state's decision to support the building of a communications technology satellite (CTS). Time and time again ensuring Canadian capability was a problem. National resources in spacecraft design included the design and manufacturing industry, and a strong government research contingent in space mechanics and electronics. The CTS would give Canadian industry important work between Telesat satellite constructions. But a purely Canadian program would cost almost $10 million more ($20 million versus $11.5 million) and NASA's expertise would be lost. Kierans maintained that US-Canada collaboration on the CTS would be the most cost-effective, while offering many benefits to Canadian industry. The cabinet once again opted for collaboration.[56]

In the space of a few years, satellite communications had contributed to the articulation of a national telecommunications policy. American attempts to establish a single global communications system as well as domestic pressures prompted the Canadian government to contemplate its interests more carefully than had previously been the case. When policy-makers did so, they discovered that the implications of the satellite went far beyond instant transmission. Depending on how it was controlled, the technology could instantly affect Canada's competitive position in leading technologies and imperil its economic well-being. Questions of control, use, and public or private ownership thus became prominent in the debate over state policy in communications.

The danger of American cultural penetration was also a determinant in the decision to launch a domestic communications satellite. Soon, however, the federal state found itself fighting on two fronts. The cultural struggle with the Americans was matched with Quebec's desire to secure independent participation in a satellite project. This further induced the federal state to stake its claim over the new technology.

PART TWO
System Overload

5 Collapse and Surrender: Telecommunications Regulation and the Canadian Transport Commission, 1967–1975

In 1967 the beleaguered Board of Transport Commissioners was dissolved and a new agency, the Canadian Transport Commission (CTC), was established to preside over the regulation of transport and communications. The repackaging made little difference in telecommunications. In the eight years of the Transport Commission between 1967 and 1975 federal regulation of telecommunications was increasingly unable to deal with its institutional crisis. The commission's mandate coincided with a period of shift and dissonance in the relationship between the state and civil society in Canada. The late 1960s and 1970s witnessed the rise of participatory impulses and ideologies, and political demands over issues that had once been considered simply economic. The triangular relationship between regulator, monopolist, and consumer served to upset the balance between those markets of the mind.

The key point of that triangle was Bell Canada. The company's operations depended on a massive and continuous capital flow that led it by the late 1960s to seek regulatory protection to shelter its operations from inflation and market uncertainty. At one time able to get along on its own, Bell was now increasingly forced into the regulatory arena. The company's annual pilgrimages to the CTC left it vulnerable to the mercies of public hearings, and more hearings attracted more intervenors. The frequency of these hearings cast a bright light on Bell operations, which resulted in progressively greater friction between Bell, the CTC, and groups vitally affected by rate increases. The almost continuous rate hearings of the early 1970s served to politically supercharge telecom regulation and made the process virtually ungovernable. Politicization had its virtues, for example promoting discussion over the nature and direction of Canadian telecommunications. Making the process more overtly political and less administrative, however, generated more tensions between utility and regulator, utility and consumer, and regulator and consumer. What is more, once these problems assumed such direct political dimensions, political solutions were required for a system which had stopped working for all three points on the triangle. Bell was convinced that it could not get proper regulatory protection to sustain growth and modernization. Consumer groups interpreted regulatory results as proof that the system had become immune to their concerns. Commissioners begged for release, and cabinet grew restive at having to deal with a growing problem.

By the time the CRTC assumed the commission's jurisdiction over telephones, telecommunications regulation had undergone a significant transformation. Consumer groups and provincial intervenors were somewhat successful in pulling telecommunications from the insulated regulatory arena into the political realm. But that success was attenuated by the state's strong preference towards technical-administrative solutions to the problems of regulating telecommunications. All of this came at a cost: regulatory paralysis and deadlock.

THE CHANGING CONTEXT OF TELECOMMUNICATIONS REGULATION, 1966–1969

Both the federal regulatory authority and Bell operated in a shifting context in the second half of the 1960s. Although the transfer of powers to the Canadian Transport Commission changed the basis for transport regulation, legislators left telecommunications regulation substantially untouched. The new commission had to operate in a more political environment, but without the benefit of a revamped statutory mandate. For the utility, the quickening pace of technological modernization and a changing political environment compelled the company to come to terms with both. The challenge to the proposed revisions to Bell's Special Act in 1967 provides an interesting window into both the technological and political consequences the company would face in the contested atmosphere of the late 1960s.

It came as no surprise to Bell's vice-president, W.H. Cruickshank, that both his company and the country were on the verge of a transformation. His main concern was to persuade Bell and Northern Electric senior executives that the pace of change in political, social, and cultural fields was quickening.[1] Bell managers did recognize that, in this context, Bell's commanding position in Canadian economic life would prompt harder questions about its operations. Adaptation was paramount, and would require the company to position itself to harness modernity, so as not to get trampled by it.

The centrepiece of any strategy of adaptation would mean bringing the government in tune with the company's objectives. Cruickshank viewed Bell's growth as "more and more susceptible to government influence." The company's political activities needed consolidation and planning in defence of its interests. The state's awakening interest in Bell would be played out most conspicuously in the regulatory arena. The BTC's 1965 hearings on the company's level of earnings put Bell squarely in the public spotlight. That scrutiny would be only the beginning. In the meantime, the BTC would undergo an extensive remodelling. The National Transportation Act of 1966 sought to remake the regulatory regime more in tune with national objectives, at least in transportation. One royal commission had warned as early as 1961 that by failing to modernize, "regulatory agencies fall into the grave danger of being subject to the industry they are expected to regulate, and of becoming the citadel of the status quo."[2] The Commission's warnings were only half heeded. When the act dissolved the BTC and created the CTC, the regulation of telecommunications passed unchanged from board to commission. The CTC would have a number of committees to deal with railway, air, water, motor, and commodity pipeline transport. But, surprisingly, telecommunications matters did not rate a committee, and so would be dealt with by the Railway Transport Committee. During the bill's hearings in the Standing Committee on Transport and Communications, there was hardly a reference to the lack of any special provision for telecommunications. The Act was significant not for what it contained, but for what it had left out.

An early manifestation of the politically charged atmosphere Bell had anticipated came with the introduction in 1967–68 of new legislation to alter the company's existing charter. As part of its plan to protect itself from nosy governments, Bell sought an increase in its capitalization to $1.75 billion as well as the power to issue stock without regulatory oversight. Most significantly, the company sought to include all telecommunications activities in its statutory mandate. "In no previous 10-year period," Bell suggested, "has the Company experienced a pace of technological change which has produced such a variety of new offerings designed to improve efficiency, productivity and convenience and meeting increasingly complex communications needs whether in the fields of voice, data or facsimile."[3] As a result, the company sought broad powers to use and adapt any means of communication whatsoever in its own interests.

The introduction of what might otherwise have been a housekeeping bill transformed itself into a lightning rod for discontent with Bell's telecommunications operations. When Bill finally made its way to the Standing Committee on Transport and Communications, its opponents were prepared. The company's bid to gain the power to issue shares without regulatory approval paled in comparison to efforts to extend its activities to anything defined as telecommunications activity. The legislation would have serious consequences for regulation.

The line-up of witnesses gave some indication of the trouble the bill would encounter. Among the more prominent witnesses against the legislation was Industrial Wire & Cable (IW&C). The company had tried and failed to get the BTC to agree to restrict the Bell-Northern relationship in 1963, and wanted to ensure that its views on the matter were restated. It took direct aim at Bell's perceived grab for power by warning that Bell intended to "stand athwart the input and output of all transmitted data" – a prospect which could not be allowed to happen.[4]

Who would supervise this institutional mammoth? The CTC made easy prey for Bell's "hordes of experts and specialists and its unlimited resources." At the very least, telecommunications should have its own special committee, and protect the public interest. IW&C had no objection to the ownership and control of the "electronic highways" as long as their non-telephone services would become regulated. The spillover from Bell's regulated to its unregulated activities was the crux of the matter, and had to be addressed.[5]

The most damaging testimony against the Bell legislation came from DCF Systems Limited, a computer communications management consulting group. Its president, Dr H.S. Gellman, presented a brief to the committee that documented the serious difficulties their clients had had in installing equipment and setting up computer communications facilities, ranging from long delays to outright obstruction. DCF's general manager, M.V. Holt, illustrated Bell's use of its monopoly powers to offer package prices for data communications, trying to drive their competitors out of the market by drastically undercutting prices to eliminate competitive offerings. D.H.W. Henry added the weight of the Combines Investigation Branch to their concerns, arguing that CTC regulation was woefully unprepared for the corporate consequences of the changes in telecommunications.[6]

The hearings also gave veteran municipal counsel Lovell Carroll an opportunity to push his favoured solution to regulatory crisis: the public utility ombudsman. Carroll had been intervening in cases since 1950. Carroll was typically appointed as counsel opposing Bell's increases two weeks before the cases started, without benefit of staff. That problem would be readily solved, Carroll argued, with the establishment of an office representing the public and examining situations as they arose. Others argued that that was what the CTC itself was for.

Given the chorus of complaint, Bell retreated. The act required Bell to establish standards governing use of non-Bell equipment along switched and private lines, with regulatory oversight by the Canadian Transport Commission on the matter. The company would still have the power to "transmit, emit or receive" communications, but was limited to its status as a common carrier. Regulatory oversight of stock issues remained unchanged. In the process, however, the company had taken several blows. One NDP member of Parliament drew his own conclusions: "If there is anyone in the House who does not believe the Bell Telephone Company eventually will have to be nationalized, then he is living in a dream world. This company should have been grabbed long ago." The sentiment was not widely shared, but it expressed a growing unease over Bell's increasing power over the field.[7]

The scrutiny over Bell's special legislation attracted the close attention of the CTC's creator and its first president, the former minister of transport. Soon after his term began, Jack Pickersgill wished to divest the CTC of telecommunications matters. Not shy about giving his cabinet successor advice, Pickersgill wrote to Paul Hellyer in August 1968 about what could be done about a situation he himself had created. Pickersgill expressed the hope that the legislation creating the Department of Communications would also "provide for the regulation of telecommunications by some other body than the Canadian Transport Commission."[8] He added that he was quite aware as a former minister that the CTC's telecommunications responsibilities were marginalized. His present post confirmed the fact: "the telecommunications aspect of the work of the Commission cannot receive the same kind of attention given to transport." He recommended that the CTC be immediately relieved of its jurisdiction in telecommunications, "and that other arrangements be made more consonant with the importance of telecommunications."

Pickersgill's admission may have been extraordinary to the student of public administration, taught to believe that agencies wished to preserve, not give away, their power. The commission's situation, however, was exceptional. The CTC's responsibility for both an expanding telecommunications system and a massive national transportation system had taxed the agency almost beyond endurance. Divesting telecommunications would save the commission from dealing with the public attention that Bell would inevitably attract. The CTC already had its hands full dealing with truckers and railway executives.

Transport's deputy minister, J.R. Baldwin, responded to the suggestion by calling attention to the studies that had begun over the revamping of telecommunications regulation and operation. The establishment of the new Department of Communications was as far as the federal government was prepared to go. By absorbing the Telecommunications and Electronics Branch of Transport the new department could call upon a base of telecommunications expertise. Moreover, responsibility for telecommunications policy would also reside in the new department. One of its first acts would be to bring unregulated private wire services into the sphere of CTC regulation by amending the Railway Act in 1969. In the early to mid-1970s, the DOC would also absorb public discontent over the CTC's handling of telecommunications, even though the department retained no direct oversight for the operations of the Canadian Transport Commission. That oversight still resided in Transport, as did the appointment of commissioners. The diffuse nature of federal authority in telecommunications regulation gave the commission a wide latitude in the application of regulation. Any changes would have to await a future round of telecommunications legislation. In the meantime, the CTC still had to regulate a massive national transportation structure on the one hand, and an expanding and technologically sophisticated telecommunications network on the other.

CTC REGULATION AND BELL TELEPHONE, 1967–1972

If Bell was hoping for regulation as usual from the CTC, the tribunal's first act would have given cause for doubt. Just as the Standing Committee hearings over Bell's Special Act were in full swing in December, 1967 the CTC asked Bell to explain exactly why it was that the company's earnings had exceeded the maximum level of earnings set by the BTC in 1966. Those earnings were to fall between 6.2 per cent and 6.6 per cent; Bell had made earnings in 1967 of between 6.7 per cent and 6.8 per cent. Bell was asked to propose rate reductions accordingly. In the end, however, the commission acquiesced in the company's explanations that demand and inflation required higher earnings. The matter was closed, but not before both federal and Ontario New Democrats called on cabinet to reverse the decision, and for Bell to give back the $76 million it had extracted from subscribers.[9]

By the end of the year, Bell was back before the commission asking for general rate increases that would yield the company $83.6 million in revenues, a 14.2 per cent increase in its total revenues. Bell asked for a total rate of return on capital of 8 per cent, with the maintenance of a debt-equity ratio of 40 per cent. Growing demand, modernization, and rising construction costs drove up the requirements. Investors moreover, were looking for a hedge against inflation, so dividends had to be more than merely "reasonable." The main thrust of Bell's case was demand for service. Between 1959 and 1969, the company had added over two million telephones and handled twice as many long distance calls. The pace of system expansion generated financial requirements that could not be met, Bell argued, "simply by doing more and more of the same old things."[10] Inflationary pressures forced the company to revise its schedule of rates, and its rate of return on capital.

Bell stressed its status as a domestic bulwark against foreign ownership of communications systems, and declared that the company had "backed this belief with action." When faced with the possibility of foreign control of major telephone systems, Bell acted to purchase majority control of them, including Avalon Telephone Company in Newfoundland, Northern Telephone, New Brunswick Telephone, and Maritime Tel & Tel. What was good for Bell was also good for Canadian communications.

Pressures from within the provinces of Ontario and Quebec led to their participation in the rate case. In the postwar period, the task of opposing Bell fell to municipalities or municipal associations. But as Lovell Carroll had made clear both in the BTC's public hearings and before parliamentary committees, the fight had gone out of the municipalities. Their collective ability to mount opposition to rate increases was in inverse proportion to the skyrocketing cost of proper legal preparation. The thrust of the municipal arguments had been based as much on common sense and practical business judgment as on abstruse figures. The municipalities would still be represented, but the primary task of representing their constituents would fall to the provinces, whose resources were more equal to the task of effective intervention.

Hearings on the application began in May 1969 with the parties laying out their evidence for and against the rate increases. The company continued to stress the great demand for its services, and innovation that stimulated even further growth. Robert Scrivener, Bell's vice-president, used the example of direct distance dialing (DDD) as a good indication of the interaction between demand and innovation. Innovations in dial service and automatic switching created the need for call-connecting equipment with higher speeds and greater capabilities. Since the service had been introduced in Toronto in the spring of 1958, 85 per cent of Bell's customers could dial over eight million telephones in Canada, and one hundred million in the United States. The management of that complex interaction should be left to the company. As A.J. de Grandpré argued, "The Bell Company relationship to its customers and shareholders ... is very similar ... to that of a guardian who administers the assets of an infant." Although it was unlikely that a custody battle would result, the infant was beginning to have ideas of her own.[11]

The Canadian Federation of Mayors and Municipalities had a simpler solution: Bell should confine itself more to basic telephone service, lessening their additional capital. In the meantime, basic service in rural areas suffered while Bell spent money on non-essential services in urban areas. Since Bell consistently made more than its permitted earnings set in 1966, the company should be denied its application. Carroll suggested that the CFMM was satisfied with the rate base as it stood. His concern, as it had been for twenty years, was with the extravagance of the construction program and holding the line on rates. His clients were prepared, however, to acquiesce to a small increase in the rate of return. Whatever the outcome, provincial involvement in the regulatory process signified a real shift in preparation and tactics from municipal strategies.[12]

Ontario's first intervention concentrated on Bell's regulated versus its non-regulated activities, and argued for a more rigorous accounting of costs to determine more precisely where its revenue was going and what the proper tariffs should be for the company's service offerings. Under its existing system, Bell was unable to allocate revenues and expenses between its regulated and unregulated services; as a result, the CTC set up the Telecommunications Costs Inquiry. The regulatory dynamics of the postwar period would now be altered, since the more powerful and articulate interventions from governments were more equal to the task of critical examination of the company's applications.

In September 1969 the CTC handed down its decision, which allowed Bell only $27.5 million in the revised rate of return, about one-third of the $83.6 million asked for in the original application. The debt ratio was left at 47.1 per cent, and the rate of return was pegged at 7.3 per cent. Commissioners Taschereau, Kirk, and Lafferty took the view that growth would continue, even if the temporary deceleration in the economy occurred.

The CTC decision signalled a sea-change in the direction of telecommunications regulation. From Bell's point of view, the regulatory climate had turned much cooler, and certainly less tolerant of its pleadings over insufficient earnings. From 1966, the company had consistently overshot its permitted rate of return, and managed to increase rates for two-fifths of its exchanges. The commission was undoubtedly aware of Bell's recent performance by the parliamentary hearings on Bell's Special Act the year before. But perhaps most significant was the intervention of the provinces in opposition to Bell's application. Though relatively mild in comparison to their subsequent interventions, provincial representation allowed the commission more opportunity to depart from fomer decisions that had been more friendly to the company. Even the speed at which they arrived at the decision – two months – indicated a departure from the old order of regulation.

Bell management viewed the $27.5 million decision as an unrelieved disaster. Plans were made for reductions in the construction program and hirings, as well as in meeting demand for telephone service. The decision had not questioned the fundamentals of Bell's case. Rather, the company argued that Bell had enough revenue to achieve their construction and modernization targets, but its strategy had been to concentrate on the implementation of large-scale technological changes. Direct Distance Dialing was only the most conspicuous among a host of expanding improvements. In the late 1960s the company began to redirect its efforts toward harnessing that technology to capture the consumer's discretionary dollar. Once basic telephone service had been provided, Bell could keep the subscriber on the line offering DDD, long distance ease, coloured phones, and extensions, along with many other services. The reorientation towards enhanced services (above basic services) required capital expansions that brought Bell into more frequent contact with both the financial community and the regulator.[13]

The company was back before the commission in 1970, a few months after the previous judgment. Bell counsel zeroed in on the subject that gave them the most trouble: the construction program. In the nine months since the 1969 judgment, construction program estimates had risen substantially. Rural service was deteriorating. Extended area service was in very high demand, especially in areas where changes in municipal organization extended local boundaries. The company went so far as to predict delays of several years in service upgrading if their application was refused.[14]

Municipal and provincial intervenors led off their critique by warning of the inflationary effects of "instant rate-making." A rate increase would "add fuel to the general inflation fire." Ontario counsel Robin Scott reminded the commissioners that only a short time had elapsed since Bell's previous round of increases. If Bell kept knocking at the CTC's door every day, it would only create trouble for the regulator.

The 1970 judgment gave increases of 6.25 per cent in all categories of exchange telephone service with the important exceptions of essential residential or business telephone services, which would increase by a general 3.75 per cent. Despite the increases, Bell had advertised that the cost of long distance was going down, and this was where the trouble for both company and commission started. The problem centred around the termination of a popular night economy plan with a new schedule of slightly higher rates. Despite protests, the changes were pushed through. Bell's strategy seemed to be geared to recovering the regulatory reversal of 1969 on a piecemeal basis.[15]

If Bell's long distance rate changes came to grief without public hearings, the company could expect serious trouble over its general application for rate increases in November 1971. The frequency and notoriety of rate hearings led the CTC to establish a separate committee for telecommunications. As Jack Pickersgill noted, he had become convinced that the burden on the Railway Transport Committee was "far too great."[16] Guy Roberge, vice-president of the commission, would preside as chairman, along with commissioner Frank Lafferty and the first woman appointee, commissioner Anne H. Carver.

The company cited the usual challenges posed by increasing demand and modernization, and predicted the usual consequences that would result from any failure to grant the revised schedule of rates. "Financing at minimum cost cannot be accomplished," Bell reasoned, "on a hand-to-mouth basis." That state of affairs could not continue without damage to Bell's ability to maintain its financial health. Bell's financial health care bill amounted to an additional $78.1 million for 1972 distributed across several categories, which would result in a rate or return of between 8.2 per cent and 9 per cent.

The combination of long distance rate hikes in the summer and a renewed application in the fall was too much for some to pass up. In many ways, Bell was the perfect target. The company was a monopoly operating a utility with a reputation for occasional heavy-handedness; its product was also ubiquitous; and, accurately or not, the telephone company represented the acme of extravagance and largesse. NDP leader David Lewis led the attack in November 1971 by demanding that Robert Stanbury, the minister of communications, nullify Bell's application before the matter had been heard by the CTC.[17] Its ally in the fight against Bell, the *Toronto Sun*, echoed NDP opposition by adding rhetorical fuel to the fire: "Bell," the paper announced, "is displaying all the characteristics of a velvet-voiced villain, oozing platitudes and propaganda while turning the financial thumb screws tighter and tighter on taxpayers."[18]

When the hearings finally got under way in March 1972, Bell opened its case by emphasizing how much the company had been burdened with the mantle of leadership in uncertain economic times. A steady rise in population and inflationary pressures made the job of implementing technological change that much harder. Bell's chief witness on the economy, W.A. Beckett, explained that Bell management had to plan for, not merely cope with, technological change. Beckett saw social change in the guise of new demands on communications providers as an important challenge to Bell. Utilities like Bell were not greedy: they just suffered from inadequate levels of earnings. Relief had to be given.

The interventions of Ontario and Quebec were much more comprehensive and meticulously documented than those of their municipal collaborators. But the two provinces' strategies were substantially different. Ontario hoped to force Bell into detailing costs in order to determine how much was actually required to maintain reasonable service. By contrast, Quebec's objectives were more directly political, shaped by its conviction that its "mission souveraine" had to include every means at its disposal to direct social and economic development in the province. In the meantime, provincial counsel argued that Bell should not receive special treatment, and should scale down its extravagant construction program and concentrate on the essentials.[19]

The CTC handed down its decision in mid-May of 1972. The decision was bound to be disappointing for the intervenors. The commission allowed an 8.2 per cent rate of return on total capital and increases of $47.2 million. It urged the company to exercise restraint in its construction program, which constituted the most important element in Bell's revenue requirement. But the wish for restraint was coupled with an equal hope that any cuts would not jeopardize the quality of service or the "Company's ability to meet reasonable and normal demands for service." The commission was convinced, at least partially, by the arguments of the intervenors that Bell's revenue forecasts were too conservative, especially considering the optimistic testimony about Canadian economic performance by Bell's own witnesses.[20]

The commission approved Bell's plans for replacing the incremental exchange group plan with the "weighted factor plan." Here, the provinces cautioned the CTC that upgrouping in the Montreal, Toronto, and Hamilton areas would eventually mean a 10 per cent increase in rates; in Windsor and Quebec, increases could be double that. The commission amended the plan, which was to be implemented in three stages. To Bell management, the decision would do little to alleviate the inflationary pressures brought to bear on the company. It was becoming more and more necessary to plan rate applications well ahead of time, since revenue requirements would compel the company to go to the commission annually.[21]

A month before the 1972 decision was handed down, Robert Scrivener had met with Robert Stanbury and complained about the CTC's attitude and warned of the consequences if Bell could not fulfil its revenue requirement. Perhaps indicating his exasperation with the CTC, Scrivener wondered whether it would not be better for "telephone services ... to be regulated primarily at the provincial level, with the federal authority being responsible for ensuring that national and international needs are attended to." In this scenario, Scrivener would leave technical and operating standards to the provinces as well as inter-regional rates.[22] Scrivener told Stanbury that he had come to this conclusion after extensive discussion with Quebec and Ontario officials, who impressed upon him the need for telephone development to reflect their own social and economic objectives. This was almost certainly a strategic declaration on Bell's part. As a company with a federal charter, strengthening federal authority in matters of national telecommunications was not likely to harm Bell. On the other hand, submitting its operations to provincial control was a leap of faith. Given the traditionally slow pace of jurisdictional transfer as well as Ottawa's understandable reluctance to give up its authority over a major corporate entity, Scrivener's discussion was likely meant to serve notice that Bell was prepared to examine all its options. This included allying the company with the provinces in the growing jurisdictional dispute over communications in order to bring pressure to bear over Bell's rates of return.

AN ARENA OF CONFLICT:
TELECOM REGULATION, 1972–1974

Between 1972 and 1974, pressures for meaningful consumer participation, the depredations of inflation, and the controversial nature of CTC decisions in telephone rate cases resulted in a political crisis for both the regulatory tribunal and the government. It also engendered serious doubts about the CTC's ability to govern an increasingly unruly process. The inflexibility of regulation made it improbable that the commission could ever accommodate the conflicting demands of the state and provincial governments and consumer groups on the one hand, and the necessity of preserving the financial health of the company on the other.

The outcome of Bell's long-term regulatory strategy surfaced in the fall of 1972 when the company filed two applications with the commission. Application A sought an order approving revisions to the company's tariffs to be effective in January 1973 and yielding $36 million of additional revenues. Application B sought rate increases beginning in January 1974, adding up to $85.2 million in new revenues. Counsel for Ontario and Quebec sought to have the applications joined in one big hearing, but failed. Bell was destined to come under sustained regulatory and public scrutiny, likely over eighteen months. This way the company could plan its attack, and perhaps even go on the offensive. Bell's case rested on management's conviction that the company had to attract financing for the rapid pace of its construction program. If measured by total telephones, Bell's growth was substantial both in absolute and relative terms. Telephones in service from 1959 to 1975 went from 3.3 million to 7.9 million, an increase of over 88 per cent, as table 17 illustrates. External financing was simply not enough. Rate increases would fuel profits and dividends, and in turn keep the construction program on a firm basis.[23] The threat of cutbacks might force the CTC to deliver the increases.

The company's applications unleashed a torrent of protest and resulted in a protracted battle for position between Bell, the CTC, and the intervenors. Apart from representation from Ontario and Quebec, the commission heard from a score of individuals and groups, ranging from labour unions to aboriginal associations. Both applications A and B sharpened the focus on the commission and highlighted its ability to handle telecommunications issues, which were becoming more and more controversial. Partly as a result of the tense climate prevailing at regulatory hearings, the CTC could anticipate more organized resistance.

The growing discontent did not take long to produce reverberations in the House of Commons. New Democrats in the House used the occasion to press for public ownership. Even if that message appealed to a minority, the question of excessive profits and consumer gouging would have a much wider audience. "The effect of Bell's A plus B theorem," the NDP member Terry Grier announced, "will be a 100 per cent increase in most general service charges, 100 per cent in pay telephone calls and residential phone tariffs will increase by around 50 cents a month." At the same time, profits were over $1 billion between 1962 and 1972. The Progressive Conservatives took the position that, as Bell Canada's largest customer, the federal government should ensure that the taxpayer did not pay more than necessary for telecommunications services. Since the state's relationship with Bell was "almost familial," as Perrin Beatty explained, the government had the ability to put an end to the massive tax write-offs that reduced Bell's taxable profits.[24]

Telecommunications regulation was not an open process. Consumer or special interest groups would find participation in public hearings an expensive proposition. To be fair, the problem of public participation in the regulatory process was relatively new. Until the 1970s, consumers, with rare exceptions, were satisfied with municipal or provincial representation before the regulatory board. Even access to hearings transcripts was difficult to secure and required cabinet intervention.

Such barriers were becoming a source of worry to the government as well. The regulatory process served to demarcate the political from the administrative and insulate policy-makers from the direct pressures of regulating rates. Bell's yearly pilgrimages to the commission had sensitized cost-conscious consumers to telecommunications increases. Pressures exerted on the CTC would eventually force the government to tackle the issue. The piles of letters the DOC received and those published daily in the general press were clearly indicative of the mounting level of public dissatisfaction with Bell. More galling to the public was the increase in the company's profits, yet Bell was returning to the CTC a mere six months after having received a sizable increase.

One DOC official offered the following comments:

The public reaction to rate increases is a profoundly hostile one. It is naturally a visceral rather than a sophisticated and urbanely reasoned one under a veneer of legal-fiscal economic double talk. This essentially "gut" hostility is sort of [an] unshakeable common-sense reaction ... There is no clearly visible change in services provided – be it in functional or even in "Chrome grill terms" – in the telephone sector to explain year to year rate advances. The hostility is focussed on the one clearly visible target – Bell.[25]

Public reaction had reached the boiling point and something had to be done about it. The solution might lie in granting an immediate increase, then opening up a score of informal hearings with "maximum and uncomplicated access" to make decisions better understood, and even accepted.

The benefits of opening up the process would provide relief along several fronts. First, the proposed reforms would divert attention from the impression of close ties between the CTC and the carriers by demonstrating that the commission had individual interests at heart. Second, it would take the pressure off a government that was trying to manage a minority Parliament. Holding the balance of power, the NDP was sure to force the issue in the House. Third, it would undercut provincial contentions that decentralization of control over telecommunications undertakings was necessary on the pretext that provincial governments were closer to the interests of the consumer. Lastly, the reforms would provide the commission with the subscriber's view, something not frequently heard in regulatory hearings.[26]

The solutions offered by DOC officials such as Allan Gotlieb and J.W. Halina were sound schemes designed to rescue regulation from losing its legitimacy and to spare the state from a growing problem. The DOC plan had the virtue of simplicity while preserving the probability of giving the company the additional revenue required for the financing of the construction program. The plan would also have the value of shoring up a vulnerable quarter of federal management.

These suggestions reached the new chairman of the Canadian Transport Commission in late November 1972. Edgar Benson regarded the suggestions as a violation of commission autonomy. Any interference on the part of the DOC might violate the boundaries between agency and government. Allan Gotlieb did not think so, "since the matters referred to are essentially procedural and do not relate to the particular merits of the rate application."[27] In any event, the final authority rested with the commission, not the department, however many bright ideas its officials may have had.

The CTC was not nearly as concerned as the DOC about the situation. In its judgment of March 1973, the commissioners wrote that they had been "aware of the concern in some quarters" about the ability of individuals to make their concerns heard in the regulatory tribunal. The CTC was having none of it, noting that much of the consuming public's comment on these rate applications indicated a "lack of understanding of the statutory functions given to the Canadian Transport Commission." In the commission's view, the public interest would be served when rates were just and reasonable while at the same time allowing the carrier to earn sufficient revenue to keep the company healthy. The matter was settled, but only temporarily, since the pressures behind the discontent had not been satisfied.

Consumer pressure was now taking an increasingly organized form. By the 1960s, the high levels of use had enhanced the position of the individual as a unit of consumption. This focus produced a series of attempts to organize consumer groups in order to seek greater product information, and to articulate concerns about products and prices. As table 18 shows, personal expenditures on goods and services exhibited strong growth in the postwar period, and kept its momentum well into the 1970s.

The consumer movement also derived its impetus from shifting conceptions of the social responsibilities of business. This meant increasing scrutiny of new products and ensuring proper pricing. Most of all, consumers were more liable to act when rising prices threatened the stability of their purchasing power. Consumer prices experienced increased upward pressure between 1966 and 1969, and especially after 1971. The average rate of increase after 1971 was over 6 per cent, double the rate between 1963 and 1969. The accelerated rate of decline in purchasing power after 1970 is so pronounced that by 1975, one 1971 dollar was worth only 71 cents. Such a noticeable decline would undoubtedly create consternation.

In Canada, consumer groups soon recognized that appealing to the state for remedy meant dealing with a panoply of regulatory tribunals. State regulation was more adept at incorporating political priorities than the unfettered market. Clearly, a higher level of organization was needed to represent the interest of the consumer effectively. Federal policy-makers had recognized the importance of better representation of the consumer by the creation of the Department of Consumer and Corporate Affairs in 1968. The legislation also created the Canadian Consumer Council to advise the minister on matters of consumer interest. The ccc was charged with improving the "quality and quantity of the dialogue between producer and consumer" to effect the smoother functioning of the mixed-market economy. That meant considering the "frontiers of consumerism" where problems existed, but solutions not yet found.[29]

The ccc's second chairman, Harold Buchwald, reported in 1972 that the council had made it its business to fight for representation of consumer interests on regulatory boards. He castigated regulatory tribunals and marketing boards who had not taken into consideration the consumer perspective in their deliberations.

In June 1973 the council published a *Report on the Consumer Interests in Regulatory Boards and Agencies*. The purpose of the study was to determine whether the consumer interest was being considered at both the federal and provincial levels of government. The findings were not encouraging. The council held the sweeping view that regulatory tribunals in general had become captured by the industries which they were intended to regulate. The report called for extensive changes to the statutory provisions of regulatory boards to allow consumer standing. With regard to the ctc, the council concluded that the regulatory culture was still bureaucratic and professional in its orientation, and especially unfriendly to interventions from consumers.[30]

Bell's treatment of its clients more as consumers of communications and less as utility subscribers emphasized the trend to consumerism in telecommunications. Table 19 shows the continuous upward trend of local and long distance calling between 1967 and 1975. Along with provincial governments, consumer groups themselves would also have to intervene on their own behalf to ensure the breadth of consumer perspectives be represented.

The revitalized defence of the consumer interest posed a difficult problem for telecommunications regulation. Economic and technical change had begun to unsettle the CTC's rate-making process; responding to consumer demands merely exacerbated matters. Accommodation and mediation would have to be achieved in a more contested environment. One activist envisioned the movement for "national citizen participation" as entering a new and "more dangerous" phase – one that would stimulate serious changes in the relationship between the citizen and the state, and between the citizen and the market.[31] The inclusion of more and more interested groups threw off the equilibrium within regulation.

The Consumers' Association of Canada was the only nationally organized body representing consumers across the country and as such had an important role to play. Originally an outgrowth of the women's section of the Wartime Prices and Trade Board, the CAC came to view the new consumer consciousness as "an almost unconscious revolt by consumers against the values of our mass consumption society." Those values included putting the "space race ahead of the human race," and fostering environments which destroyed "creativity and satisfaction." Consumerism in its extreme form, Maryon Brechin of the CAC told the Canadian Telecommunications Carriers Association, resulted in radical and violent protests that were springing up around the world.[32]

Closer to home, Brechin argued that the public interest in telecommunications regulation was so vague as to be useless to the CTC and related bodies. Regulatory proceedings tended to be excessively technical and costly. "Too often hearings become the jousting ground for government and industry interests," Brechin suggested, "with the consumer a muzzled spectator on the sideline." Communications expenditure was about one-third of the household operational budget, excluding heating and lighting. Given what was at stake, it came as no surprise that the alienation of the Canadian consumer from the regulatory process ran deep.

The CAC's own troubled attempts at defending the consumer at hearings reflected the problem. The association spent a great deal of time and much of its limited resources on intervening, only to find that all but $4 million of Bell's requested increase would be granted. It was small comfort that the reduction occurred in areas the CAC had emphasized: charges in pay telephones, for example. Frustration was the general reaction.[33]

One solution which had been advocated as early as 1965 was the establishment of a consumer advocate or ombudsman. The idea was given more thought in 1972 when the Canadian Consumer Council published a study by Michael Trebilcock entitled *The Case for a Consumer Advocate*. Trebilcock cited the 1972 Bell rate hearing as a typical example of where an advocate was needed. In that hearing, "the only direct consumer participation," Trebilcock noted, "was a Quebec farmer who asked an occasional question and a woman who was allowed to speak for half an hour at the end of the hearings." The process was simply not designed to include the ordinary consumer. The solution lay in the "enfranchisement of all affected interests." The continued legitimacy of the regulatory process was in play. The current random and scattered representation of consumer interests contributed to the imbalance in the regulatory process.

The character of the hearings had expanded significantly to include a score of intervenors representing consumer and labour groups. Groups such as Action Bell Canada and Native Marathon Dreams were loose coalitions of individuals and, like virtually all the intervenors, claimed to speak for the protection of the consumer interest. In most cases, intervention meant opposition to increases. Native Marathon Dreams undoubtedly spoke for many in criticizing the position of the commission which had "merely rubber-stamped previous applications by Bell for price hikes."[34]

Rate hearings attracted a barrage of demands, often not strictly related to general rates or service. One group called the Individual Information Institute demanded that Bell publish a directory of "public information" in large population areas. Their brief noted that "people will come to identify and use the telephone as an information source when the telephone is identified as such."[35] The rate hearings had not only become a focal point for dissent, but also attracted activists for a variety of causes.

As they had done before, the intervening provinces mounted a comprehensive examination of Bell's evidence. The conclusions were familiar. Quebec counsel zeroed in on the construction program. That program had "toujours servie de prétextes à l'augmentation des taxes," and risked becoming a perpetual justification for whatever increases Bell saw fit to ask for.[36] What galled the main intervenors was that Bell's own policies had deliberately stimulated the need for greater construction expenditure through promotions such as "Bargain

Month" which provided discounts and rebates for connection and rental charges.

In contrast to Quebec's clear rejection of the rate increases, Ontario's position was more tentative. Provincial counsel D.W. Burtnick continued the argument over the construction program and questioned the motivation behind Bargain Month. In general, however, Ontario was more concerned with highlighting evidence and less with drawing dramatic conclusions about Bell motives.[37]

Bell responded to the interventions with a certain exasperation. Bell counsel Ernie Saunders summed up the besieged feeling of the applicant. Any attempt by the provinces, Saunders argued, to interfere in the management decisions of the company had no place in the hearing. Saunders was blunt: "If the Provinces of Ontario and Quebec desire to have the right to interfere in the management of Bell Canada ... then they should be prepared to assume the responsibility that goes with that right." In the meantime, the result of succumbing to the arguments of the intervenors would mean serious problems in maintaining the level of service to which Bell's customers were accustomed. Bell could not keep up with both demand and modernization.[38]

Bell argued that management had "little flexibility" in the timing of capital expenditures. Growth in existing and new services accounted for two-thirds of the construction expenditure. Unless the company were permitted to plan long-range and commit capital resources, service deterioration was an inevitability.

The prophecies of Bell counsel had their intended effect in some quarters. Department of Communications observers noted that if an increase were not granted, the company's ability to provide service would be "seriously impaired," resulting in a potential forty thousand to fifty thousand held orders for basic service and three times that for service upgrades. Both the political consequences and the effect on the telecommunications system would be significant.

In its decision of 30 March 1973, the CTC agreed with Bell that a 7.8 per cent rate of return would be adequate, but disagreed that the $36 million in additional revenue was necessary to reach the goal. The commission accorded increases that would generate an additional $29 million in revenue. On the crucial question of the construction program, the commission essentially agreed with the company over both its estimates and the importance Bell accorded to construction.[39]

The commission's decision caused an immediate political reaction. On 2 April the government found itself pressured to rescind the increases as soon as possible. The decision had put cabinet in a precarious position. Respecting the independent nature of the regulatory process was one thing; dealing with an unpopular regulatory decision in a minority Parliament was another. The CTC decision gave the opposition the platform they needed to rail against both Bell increases and Liberal inaction.

The NDP hammered home their thesis that Bell had gouged the public, that its profits were excessive, and that the company was not paying its fair share of the tax burden. Members who had assailed the old Board of Transport Commissioners in its time now pined for the days of its regulatory oversight. Bell had done an excellent job in telecommunications, it was true, but for the NDP the time for nationalization had come just the same.[40]

Tory MP Alvin Hamilton agreed with part of the diagnosis, but not the cure. Hamilton argued that the fault lay with the regulator itself. The commission had been handed an excessive amount of power, and as a result they seemed to consider themselves infallible. "Even Pope John," Hamilton quipped, "gave up the notion of infallibility twenty years ago." Bell should be made to act in the national interest, although Hamilton declined to define what precisely that was.

The government benches were quick to defend the independence of regulatory tribunals, but made little effort to defend the CTC itself. Instead, the government concentrated on the "unfortunate inadequacy" of the present regulatory structure, especially given the fusion of transportation and communications in one regulatory body. The minister of communications, Gérard Pélletier, offered token resistance to the opposition attacks by arguing that the cabinet needed more time to study the rate increases. Pélletier conceded that "circumstances have changed faster than the Commission could transform itself," but the changes suggested in the government's 1973 proposals for a national telecommunications policy would alter that. A few days later, Pélletier announced that the cabinet had ordered a temporary suspension of the CTC's decision "pending clarification" of the issues.[41]

"Neither considerations of statesmanship nor even of mere good government," wrote W.A. Wilson in the *Montreal Star*, "but only simple fear of imminent defeat in the House of Commons impelled the government to interfere with the Bell Telephone rate increases." The government had caved in, perhaps predictably, in a minority Parliament. One minister anonymously demanded to know whether "that son-of-a-bitch" Edgar Benson would ever stop embarrassing the government.

Reaction to the CTC decision forced the issue to the top of the cabinet agenda. Department of Communications officials argued privately that the commission did not provide convincing rationale for its decision, especially since there was neither any explicit discussion of Bell's construction program nor comment on the question of deferred taxes. Yet both were critical components of the company's financial situation. One week after that meeting, on the morning of 5 April, the cabinet met to consider what was becoming an issue with serious political implications.

Gérard Pélletier's advice exemplified the dilemma in which the CTC had placed the cabinet. If the government acted to rescind the rate increases, Bell would find the task of raising substantial sums of money it needed to finance its construction program more difficult. As a result, Pélletier proposed that the cabinet authorize him to approach the CTC to justify their recent ruling. Politically, the problems were greater. Neither Pélletier nor any of his officials had been given warning of the decision. To make matters worse, the CTC had ordered Bell to implement the changes soon after the decision had been handed down. The request for further information from the commission would allow the strong opposition in the House to dissipate while allowing the government time to consider its options.[42]

The cabinet was split. Most agreed that the public had little confidence in the CTC. Given the impending pressures for freight rate increases, failure to do anything on telephone rate increases would result in a loss of "all credibility" in the government's attempts to dampen inflation. Minister of Transport Jean Marchand suggested that the time had come to consider influencing the management of Bell Canada, perhaps by buying shares in the utility.

On the other side of the table, Treasury Board President C.M. Drury and Finance Minister John Turner were more circumspect about even a temporary suspension of the CTC decision. Such an action, they argued, might undermine business confidence. One proposed solution was to have Bell "voluntarily" withhold the implementation of the new rates. The suggestion was rejected by the more politically astute ministers, who believed the government could not be seen depending on any voluntary corporate act of goodwill. The increases were accordingly suspended so that the cabinet could study the matter "in the light of additional information." The question of suspension had special significance, given the crisis point which regulation had reached. Just three days before the CTC decision on Bell's application, Pélletier and Herb Gray, the minister of consumer and corporate affairs, circulated a memorandum to the cabinet on the growing problem of representing the consumer in the telecommunications regulatory process.[43]

Public discontent over the frequency of rate applications concentrated cabinet attention on the CTC. The regulatory process seemed either paralysed or indifferent. The consequent void had to a limited

extent been filled by Ontario and Quebec, "which appear," one cabinet memorandum noted, "to find it politically attractive to be cast in
the role of sole protectors of the interests" of consumers. Public perception equated the CTC with the government itself, both indifferent
to the consumer interest, and both partial to the corporate interest.

The difficulties of regulation "in the public interest" – rationalizing
conflicting interests, the dangers of "over-identification" with the
regulated company – were accentuated in the regulation of telecommunications. The commission, as a court of record, had to decide on
the basis of evidence presented before it. Telecommunications carriers could afford to buy the best evidence to make their case. In the
1973 hearings alone, the prohibitive costs of mounting a credible
defence of the consumer interest prompted a host of organizations,
including the Consumers' Association, to abstain from the hearings.
The political consequences of estranging the consumer interest from
regulation were well understood around the cabinet table.

The instrument that best met cabinet concerns about credibility
and competence was the funding of a non-governmental organization to represent the consumer interest. The organization would have
to be national in scope, close to the consumer, and independent from
government. The search led to the doors of the Consumers' Association.[42] The support would come from an unconditional grant of
$25,000, to come out of both DOC and DCCA budgets, ensuring that
the political concerns of the two departments would be "recorded"
by the CAC.

Despite the potential risks the grant might produce, especially over
the association's ability to represent specific regional consumer concerns, the advantages overruled any pitfalls. The intended outcome
would be to re-establish the link between the federal government and
the interests of the telecommunications user. Any provincial political
advantage in intervening before the CTC would be weakened, especially the provinces' role as champion of the underdog. The plan
would have the added benefit of undermining provincial claims that,
in the name of the consumer, jurisdiction should be provincialized,
where regulation would be more responsive.

In the month that followed, DOC officials hastened to propose a
further solution to the problem by proposing the immediate transfer
of telecommunications regulation to the Canadian Radio and Television Commission (CRTC). The broader plan for eventual transfer,
possibly in 1975, had already been approved. Pélletier now argued
that the transfer should be enacted before the coming summer
recess. The public had lost confidence in the way the federal agency

regulated the telephone industry; cabinet suspension of the rate increases indicated the government had come to a similar conclusion.

The proposed transfer to the CRTC would involve no change of powers, duties, or functions that resided with the CTC. The department's proposals for extended regulatory powers and functions would have to await a second phase of communications legislation following federal-provincial consultations.[45] The only way to restore regulation's credibility with consumers and investors was by transfer from CTC to CRTC. The cabinet had to act, or "the present unsatisfactory situation may continue for a considerable period of time." Timing was everything: if legislation was not passed expediting the transfer to the CRTC, the danger of destabilizing telecommunications regulation was real. The plan would also cause possible harm to broadcasting regulation resulting from regulatory uncertainty.

It was a tall order to fill. The DOC was asking the cabinet to approve the immediate drafting of new legislation to be passed through a minority Parliament. The department also was insisting upon an immediate build-up of financial and man-year resources to aid the new commission. In addition, not all of Pélletier's colleagues were convinced that the cure was not worse than the disease.

Pélletier's recommendation of radical surgery met with coolness from the Priorities and Planning Committee of Cabinet. The committee instructed the minister of justice to determine the feasibility of the proposed CTC-CRTC transfer by order-in-council. Failing that, the strategy would revert to consultation with opposition parties for indication whether the legislation would receive quick passage. The cabinet meeting two weeks later exposed the weakness of the strategy. Otto Lang, now the minister of justice, informed his colleagues that legislation was indeed necessary, and that the NDP would not oppose the legislation. The Progressive Conservatives were another matter, however, and refused to hold any "meaningful discussions" with the government over the transfer.[46]

In the meantime, Pélletier reported that CTC chairman Edgar Benson had called on him to express serious reservations over the action. The proposed transfer, Benson maintained, would thoroughly demoralize the commission staff. Later, Benson decided to increase the Telecommunications Committee staff. His road-to-Damascus conversion had the effect of undercutting the DOC's strategy of transfer to the CRTC. The move to increase telecommunications personnel would, Pélletier believed, "lead to serious difficulties," even though he himself had suggested building up the staff as a second (although not preferred) alternative to transfer. Pélletier's plan would include input from both the Treasury Board and the DOC, but Benson's action shut that door. Cabinet eventually agreed in principle on the desirability of the CTC-CRTC transfer, but not immediately.[47] The department's strategy was effectively emasculated; Pélletier had asked for quick legislative action before June. In this case, the cabinet illustrated its desire to avoid the perils of rushing legislation through a minority Parliament. It also did not share Pélletier's sense of urgency over the issue.

One month later, the cabinet finally revised the CTC decision on the Bell "A" project, denying the proposed 50 per cent increase in service charges. It also requested that the commission examine the social impact of any contemplated increases for residential installation charges, especially on low-income subscribers. Cabinet also found the construction program to be inadequately justified by the company, and the information provided either too general or too specific to be useful. As a result, the government urged the commission to order Bell to "provide an analysis of the major construction programs which it proposes to undertake." The government review of the commission's decision was mild in comparison to the stronger action originally contemplated. Nonetheless, the cabinet's directive would provide some sort of guidance to the commission, especially considering that the second part of Bell's application was expected to be heard in the fall and winter.

Taken together, the events surrounding the CTC decision from March to June demonstrated the extent of the crisis of confidence in the commission. The fears of DOC officials, that even minimal credibility could not be recovered without substantial changes to the architecture of regulation, were well founded. Departmental plans and cabinet discussions also revealed the serious consequences generated by the rigidity of CTC regulation. By 1973, the exclusion of consumers from the process resulted in a political predicament for the state. Yet it was a predicament partly of its own design. Since the mid-1960s, CTC regulation over telecommunications had been questioned, but it was only at the crisis point that the government had decided to act.

The outcome also illustrated the effective limits on cabinet attempts to solve the crisis. Despite strong incentives to the contrary, it opted for the status quo in order to protect its survival in a minority Parliament. The modest variance of the CTC decision gave some direction to subsequent commission dealings. It put pressure on the commission to induce Bell to come up with better analysis of the costs of the construction programme. But the lack of any strong corrective left the situation substantially unchanged, with a commission scarcely able to perform with public and investor confidence. It was in this atmosphere that the CTC began consideration of Bell's second application in the winter of 1974.

A LEGITIMACY IN DOUBT

The CTC attempted to expand the regulatory arena in light of its regulatory and political experience of the previous two years. Bell's 1974 application for higher rates provided an opportunity to implement the commission's strategy for recovering some measure of its legitimacy. By the end of the 1974 hearings, however, it was evident that once lost, the CTC's ability to reconcile rival interests could not be regained without major restructuring. Both consumer and government participants in the process remained dissatisfied that their interests had been served.

The hearings on Application B extended from the winter of 1974 into the mid-spring, comprising forty-seven days of hearings. Given the intense public scrutiny under which the hearing would operate, the commission decided to schedule hearings for Montreal and Toronto in addition to the normal Ottawa sittings. An examiner was sent to the Northwest Territories and held hearings in Frobisher Bay and Coral Harbour in February 1974.

When the hearings on Bell's second application finally got under way, the list of intervenors had grown considerably. The Commission had even asked the Department of Consumer and Corporate Affairs to provide funding for legal assistance to intervenors, arguing that no money had been provided by Parliament for that purpose. In lieu of money, the CTC, with the cooperation of Consumer and Corporate Affairs, made transcripts of the hearings available in Toronto, Ottawa, Montreal, Quebec City, and Windsor. The process had now become a stage for broader political and social issues to be discussed, so the resulting demands on the CTC strained its resources and the patience of its commissioners to the limit.

The hearings began with the ostentatious withdrawal of the Greater Montreal Anti-Poverty Co-ordinating Committee, since neither the minister of communications nor the CTC consented to a study on the effects of telephone rate increases on the poor. Ken Rubin of a consumer group called Action Bell Canada echoed their remarks. Bell could not be allowed to set its own priorities when they were so clearly detrimental to low-income poor, rural folk, and northerners.[48] Ontario and Quebec announced their intentions to test the application against their regular standards of whether they were just and reasonable. Most importantly, they would speak for the consumers of their respective provinces.

Bell opened its case by stressing the serious effects that the high rate of inflation was having on the company's operations. Its financial flexibility in that circumstance, Bell's witnesses claimed, was deteriorating as well. A heavy reliance on debt financing reduced the company's ability to raise capital. A rate of return of not less than 12 per cent was required to stabilize matters.

The hearings sifted through intense testimony about most aspects of Bell operations. The company was closely questioned on its construction program, with neither side able to declare an advantage in the hearings. Bell's Ernie Saunders wryly noted in opening Bell's final arguments that he might be "sounding like a broken record," citing mysterious economic forces that led to material shortages, inflation, increases in the costs of money, and disarray in the financial markets. But those were the facts. Saunders expressed Bell's exasperation with the marathon hearings and the numerous intervenors. "It is time," he told the commission, "that the public at large, if they intend to interest themselves in rate applications, made some effort to understand the problems faced by Bell Canada and the problems that are placed before the Commission for decision."[49] They should not be too interested, however, since telecommunications matters were exceedingly complex. That is why, Saunders argued, the CTC existed – to have experts adjudicate these matters and make decisions in the public interest. There was such a thing as too much participation.

The opposition to the rate increases demonstrated a spectrum of views about how the utility should be run, and what principles should apply in setting rates. The thrust of Quebec's intervention was to establish the socio-economic importance of the telephone. As such, the province stressed the necessity of measuring the need for increases against the necessity to maintain basic telephone service. "Nous soumettons," Ross Goodwin argued, "que le principe et le critère de base devraient tendre à faire porter toute augmentation de tarifs sur ceux qui bénéficent plus amplement du service et sur ceux qui exigent des services que l'on peut qualifier 'de luxe' et non essentiels."[50]

The experience of Ontario municipalities in Bell rate cases had left them exasperated. The Association of Municipalities of Ontario's Special Committee on Bell Canada, under the chairmanship of Niagara councillor Mel Swart, recommended after the end of the 1974 hearings that the association should concentrate instead on lobbying the federal government to "more adequately and responsively regulate the telecommunications field." Municipalities had come to the view that the liability involved in representing the local consumer interests should rightly be borne by the provincial authorities. This would be their last case. The association called for both senior levels of government to "expedite the development of a more sensitive system" for the regulation of Bell which would include better protection for the consumer.

Aside from its familiar arguments about the construction program, Ontario counsel Dan Burtnick charged that Bell had pushed the growth of the telephone network beyond its basic character into new kinds of expenditures. He concluded, that as a result of deliberate policy, Bell was "not just a telephone company anymore," but something much bigger. The value-of-service principle had resulted in an all-purpose justification for whatever the company wanted to promote in its application. Changing technology and consumer demand made for climate of rapid change; the need for a closer examination of the rate base was all the more necessary. The CAC underscored those concerns.[51]

Bell's application prompted more than the usual flood of complaints about the rate increases. Increasingly submissions cited the idea that the telecommunications system had become an essential factor in the ordinary person's life. As such, consumers had to be protected from Bell's continued indifference to their needs. "Unfortunately," Ronald Cohen, director for the Centre for Public Interest Law wrote, "the CTC has developed an overly cozey [sic] relationship with the utility it is supposed to be regulating with the result that the Canadian consumer has been caught like a monkey in the middle without a hope of touching or affecting the flight of the financial football soaring over its head." The situation could not foster a sense that the public interest was being looked after.[52]

The Telecommunication Committee found itself in a quandary after the hearings. The trio of Roberge, Carver, and Lafferty must have known that regardless of the decision they handed down, challenge was certain to come. As *Time* reported in February 1974, some commission officials were understandably unhappy with the complexities of their task. The committee chairman, G.F. Lafferty, acknowledged the dissatisfaction in "some sectors of the public" about the difficulty of making their voices heard in the regulatory process – this after eight thousand pages of transcripts, 470 exhibits, 291 interrogatories to Bell, as well as 111 formal intervenors. The fact that the hearing was the biggest telecommunications rate hearing in Canadian history did not lessen the general unease with the process and outcome of the regulation.

When the commission handed down its decision, it lowered Bell's revenue requirement to $51.8 million for a total rate of return of 8.6 per cent. Generally, however, the CTC's decision supported across-the-board rate increases and higher rates of return. It clearly sided

with Bell, especially in the company's claim to take into account inflationary pressures. The awarding of the rate increases would have to
suffice while the commission and interested parties hammered out a
more permanent solution to the problem of rate adjustment in an era
of inflationary pressure. There was little to indicate that a serious re-
examination of Bell's fundamental assumptions had taken place. The
review did prompt an unprecedented level of complexity and detail
on the program, especially in Bell's modernization plans. Combined
with the cross-examinations of Ontario and Quebec, the commission
could plausibly argue that the matter had received a thorough public
airing. The CTC could also have claimed to have given moral support
to the company amidst its many detractors.[53]

The commission requested, but did not order, Bell Canada to provide for a better grade of non-urban service. The CTC agreed with
the utility that expenses other than the basic telephone plant were
important to benefit from technological advances. Nonetheless, the
commission noted that even the Bell president, Robert Scrivener,
agreed that rural plant was "absolutely inadequate" and that the
projected expenditures of $75 million would barely make a dent in
the problem. Table 20 indicates the extent of the modernization problems, especially coping with the high numbers of multi-party lines.

Groups that had energetically opposed the application appealed to
the federal government for another variance of the commission's
decision. Opposition parties and the Canadian Labour Congress
called for a variety of responses, from rescission to nationalization.
For the Greater Montreal Anti-Poverty Co-ordinating Committee,
rescission was a vindication for its withdrawal from the hearings at
the beginning. Labour and consumer groups charged that the commission was not representative of the community and that the agency
should be restructured to become more representative.[54]

This time Pélletier defended the CTC decision, despite the extraordinary profit increases the utility had posted. However, the government did take the opportunity to reject Bell's suggestion that an
automatic rate-adjustment formula be implemented. Ontario's reaction to the CTC decision was negative, perhaps reflecting the province's ambivalent position to both the specific issue of Bell rates, and
the general question about the regulatory process. This position
reflected province's divided attitude; most consumers simultaneously
desired to allow Bell to maintain a reliable communications network
while not wanting the company to get away with excessive profits.

From the beginning, Ontario's participation in the hearings stemmed
from a desire to have its own interests recognized in telecommunications matters. Those interests were not immediately clear,

however. "We must come to grips with what we really want to do with Bell Canada, or Bell Canada to do for us," MTC communications branch director W.A. Rathbun offered, "[b]ut let us make sure that the decision is made by the Government of Ontario, not by Bell or other governments, including Quebec, BC or Ottawa."[55] A good part of the strategy would be played out on the political side and would include establishing a major presence in the regulatory process.

By 1972, the largest portion of the MTC's communications branch resources was devoted to developing expertise in the federally-regulated common carriers which dominated the life of the province. The modest objectives that the Ontario government had set for itself in this regard fit nicely with its role at the commission. CTC hearings provided the opportunity for the government to act as a vigorous watch-dog in ensuring that provincial interests were fully represented. In the long term, the province could pursue a more articulate provincial role via federal-provincial negotiations on jurisdiction.[56]

More rate hearings compelled the MTC to formulate more precise regulatory objectives. Its deputy minister, A.T.C. McNab, argued before the Ontario cabinet that the province's position stemmed from two main preoccupations: protection of the users of Bell services from unreasonable increases; and more important, a positive acknowledgment of the province's dependence on Bell for the "conduct of normal business and social activities in Ontario."[52] But so long as Bell took the arrogant attitude that increases were necessary based on Bell's self-proclaimed financial base, then the provinces's duty was to oppose the company as "hard as possible."

By 1974, Ontario had become the most comprehensive intervenor in the rate cases. The strategy of pushing for more costing information from Bell had contributed to the comprehensiveness of the regulatory process. Ontario's interventions were extensive, detailed, and had put the company on the defensive in a number of areas. It had clearly succeeded in shedding light on the problems of telecommunications regulation. The exercise had given the province an opportunity to generate research in telecommunications from the provincial point of view, and develop the necessary expertise.

Regulatory intervention had not, however, advanced Ontario's general position in communications. Indeed, it was hard to define exactly what Ontario's policy objectives actually were in the field. "Simply put," Ontario political advisor Ron Atkey noted, "Ontario has not made the case to the general public why the provincial government should be involved in communications regulation in the first place, other than to talk about the need for flexibility and regional diversity." The province had to stress that policy-makers at

Queen's Park were more capable than the CTC in protecting the consumer interest. Ontario could exploit the low public esteem for the commission by stressing that provincial initiatives would be closer to the needs of both business and consumer groups. The regulatory strategy, when fused with its political undertakings, could pay off for Ontario. Even Bell could be conscripted with the promise of more profit and less bureaucratic regulation should regulation fall to the province. If this scenario did not play out by the end of 1975, however, Atkey advised that the province vacate the field and "direct its energy and resources to other important areas."[58]

Considering this, there was disagreement on how to proceed. Some had argued that a hard-line approach would benefit provincial objectives; others favoured the status quo. MTC officials, often fresh from battles with Bell in the regulatory arena, were suspicious of the company's motives in meeting with provincial officials. After one such meeting between MTC and Bell in March, 1975, one MTC official expressed his exasperation:

Each step of [Bell's] argument is bent as far as it will go without becoming a literal lie. Each method of allocating costs is chosen from the universe of possible methodologies to favour the answer Bell wants ... It is obvious that this dictates a difficult position for the staff. They are required to watch this exhibition of legerdemain knowing full well how the tricks are done but are powerless to expose them as tricks: the arguments are too complex and any sustained pursuit degenerates into bad manners ... One can only be subject to so much perversion before it begins to gag one.[59]

The official recommended that Bell be limited in its access to policy-makers. If not, the communications division should train a small staff of "propaganda analysts" to do nothing but act as truth squad to Bell's deplorable drains on extremely limited resources.

Despite the views of the MTC officials, the province opted for the status quo, preferring a continuation of its "show cause" interventions and discussion with the federal government over confrontation. Still, the steady opposition to the increases prompted Bell to view Ontario's posture as adversarial.

Instructions by the Ontario government for the 1975 Bell application were to make a moderate intervention. The cabinet was sympathetic to Bell's increasing need to meet inflation, so it would be the task of the Ontario intervenors to "appear constructive without looking vindictive." Nonetheless, Bell continued to take exception with Ontario's energetic interventions in the regulatory process. When Orland Tropea, a Bell vice-president, questioned the merits of intervention at rate cases, he argued that Ontario's position before the CTC had supported the interests of "radical groups" whose agenda was inimical to the public interest." MTC officials refrained from reacting to Bell's "literal lie" in the interests of greater harmony.

Bell strategy to short-circuit provincial opposition by putting pressure on senior government officials was evident a couple of years later. After a meeting between Bell officials and Treasurer Darcy McKeough, where Bell expressed continued concern about Ontario intervention, McKeough later wrote to the premier to see if anything could be done. "There are times," McKeough wrote, "when the system seems to run away with itself ... the lawyers at the junior end just go crazy." A way had to be found to harmonize their attitudes with those of the cabinet moderates and the general provincial interest.[60]

Clearly, then, Ontario's attitude reflected many of the frustrations generated by the regulatory process. The province's general perspective did harden between 1968 and 1974, but in the final analysis, Ontario was satisfied with intervening and bringing Bell to account. Events had unsettled policy-makers, but not enough for the province to call for radical changes.

RESCUING TELECOMMUNICATIONS REGULATION: THE AUTOMATIC RATE ADJUSTMENT FORMULA

The problem of regulatory paralysis led Bell and the CTC to introduce a scheme designed to eliminate frequent and debilitating public hearings. In light of the failure of the proposals, all the participants in the regulatory process were condemned to another round of hearings in 1975, which were destined to demonstrate once again the failure of the commission's regulation of telecommunications.

The commission's experience throughout 1973 and 1974 prompted it to discuss dispensing with lengthy and frequent public hearings. Inflationary pressure had impaired the ability of the commission to respond to adjust prices. As a result, the automatic rate adjustment formula (ARAF) was proposed. The idea was floated by Bell in 1974, and obviously caught the imagination of the weary commissioners. The plan would have the virtue of granting routine cost increases without the necessity of public hearings. The formula was aimed at reducing regulatory delay, and lead to easier and less costly financing. The consumer price index was registering sharper increases in the cost of living, and the cost of housing was under upward pressure. The commission decided to use a gross national expenditure implict price index to determine what was to be covered under the formula. This specific approach would not take into account the cost of capital, and would equally apply to each carrier under CTC jurisdiction. The ARAF would be required to have the virtues of simplicity, feasibility, compatibility, acceptability, defensibility, and would apply only to uncontrollable cost increases.

Both Bell and the CTC viewed the automatic formula as a solution to a number of intractable problems. First, the company regarded itself as particularly vulnerable to the persistently high levels of inflation. The necessity of eliminating the regulatory lag in meeting increases in uncontrollable costs would go a long way towards reducing inflationary pressures. Second, the company would receive at least partial protection from full-scale rate hearings. By substituting what Bell called "a condensed method of regulatory review and analysis," the company would escape what had become a debilitating cycle of perpetual public hearings.[61]

The CTC proposal attracted widespread industry support, especially from the Canadian Telecommunications Carriers Association (CTCA), a telephone company lobby group. The association went further than Bell was prepared to go, arguing that the inclusion of the cost of capital and income taxes should be included in the formula. The CTC plans to discuss the automatic formula received a strong caution from the federal cabinet in early September.[62]

Other reaction to the automatic formula was mixed. The CAC was unsure where to stand on the issue. The idea had some merit, but the "realities of the regulatory environment" detracted from the merit of the idea. The goal of inflation protection for an industry which was particularly vulnerable was good, but a simple rate adjustment formula might not provide adequate protection against either unforeseen costs or potential exploitation by the carriers themselves. So Ontario eventually opposed the idea of the adjustment formula. The province's principal objection was over its fears that the formula would result in bad regulation with excessive costs to the consumer, while removing the incentive to efficient operation.[63]

Quebec joined Ontario's opposition to the plan. Its position had been clear virtually from the outset of the CTC's announcement of a hearing on the matter. The province was convinced that the issue was not one properly discussed in a regulatory forum. For Jean-Paul L'Allier, the province's communications minister, it was "une question de politique qui doit être discutée au niveau ministeriel, entre le ministre fédéral et tous les ministres provinciaux."[64] It was probable that the proposal for federally regulated carriers would translate into demands for provincially regulated telephone companies as well.

The automatic formula did not see the light of day, despite strong backing from both the commission and the company. The enthusiasm with which both the CTC and Bell embraced the concept was a measure of how contentious the regulatory process had become. The formula had the virtue of restoring some of the equilibrium within regulation lost by the proliferation of intervenors and publicity. Once at least part of the regulatory process had been "repatriated" from disputed terrain, the company could expect an easier time in getting its revenue requirements. But as table 21 shows, Bell would have to explain why its total operating revenues had averaged 9.7 per cent growth between 1968 and 1975, or why the impressive growth in long distance revenues could not more adequately cover revenue shortfalls. As the industry across Canada registered a profit exceeding $404 million in 1975, it was not immediately evident to the ordinary Bell customer why the company merited automatic increases. The failure of the automatic formula proposals meant that an increasingly impatient company and an exhausted commission could expect no immediate relief from the yoke of public hearings.

By the time of the CTC's 1975 hearings on Bell's renewed application for further revision in the schedule of rates, the regulatory process had become an exercise in frustration. With the automatic rate adjustment formula stalled in its tracks, the only hope for Bell was for quick financial relief followed by a short hearing. In the first application, made in June 1975, the company sought immediate rate relief before the end of the month. Bell also recommended the implementation by August of a new system of counting telephones. The second phase of the application sought approval of rate revisions for implementation in October, 1975.[65]

During the hearing, Bell president A.J. de Grandpré warned that the consequences of cutbacks would be felt throughout Bell territory. Under cross-examination by Ontario, de Grandpré explained that he would have to strike a balance in any cutback between the "socially acceptable" projects and the profitable ones. Non-urban service would suffer the most. If the $28 million in interim increases were not granted, the result would be a $100 million reduction in the construction program in 1975. Ontario objected to the threat of construction cutbacks, noting that the "need or necessity of doing so has not been proven" in the interim case.

In late July of 1975 the CTC awarded only half of the $28 million requested by the company. Bell executives immediately began to implement a strategy of strategic cutbacks of the construction program. A.J. de Grandpré wondered how the commission could first recognize the urgency of the request, then concluded that Bell could "still somehow get by on 50 percent of what is required."[66]

The cutbacks received the immediate attention they were intended to generate. The company had targeted phone installation for maximum effect. "If the company believes its financial plight would be given attention," an editorial in the *Financial Times* commented, Bell "should hardly have been surprised when every pundit, politician, house builder and other self-servers jumped fearlessly into the fun action of whacking the old girl's backside." Bell chairman Robert Scrivener responded that "too little and too late regulation is going to mean too little and too late telephone service."

Bell reduced the size of its construction program to $803 million from $836 million. Although the company deferred some non-service expenditures, the $33 million reduction could not have been accomplished without affecting requests for new service. In particular, growth and modernization suffered the most from the cutbacks. The message of the "emergency short term measures" would be clear, but not devastating; irritating, but not radical.

Some editorialists spoke of a war between the CTC and Bell Canada – "a war," Geoffrey Stevens wrote, "in which there can be only one victim, the consumer."[67] The company complained that its rates were consistently lower than the rate of inflation, resulting in major problems for the company. But the temporary problems of inflation were but symptomatic of a larger malaise that featured regulatory lag and frequency of applications.

The company was positioning itself for the next round by issuing periodic warnings that persistent denials by the CTC would result in drastic cuts in service. Bell had requested $110.3 million. "At some point," Robert Scrivener suggested, "there has to be a confrontation." Scrivener viewed the action as the culmination of a few years of regulation where the commission's spirit was willing, but its decisions were weak. Bell began to scale back its operations in strategic areas, suspending student employment, freezing hiring, and reducing overtime to an essential minimum.

Bell's final application was heard in November 1975. By that time, the problem of inflation had not only prompted the government into wage and price controls, but also elicited a stream of concerned groups at the CTC's public hearings. The problem of automatic increases was, for the moment, dealt with by the imposition of global wage and price controls.

The NDP MP, Cyril Symes, appeared before the commission to complain of the utility's "flagrant insensitivity" to the real needs of the consumers of Ontario and Quebec. He called once again on the government to nationalize the utility. Symes's accusations against Bell drew a sharp response from Bell's chief counsel, E.E. Saunders. He suggested that the NDP's submission was valueless and politically motivated. Saunders argued that high rates of inflation were having a devastating effect on publicly regulated utilities. The dramatic change in the company's first quarter earnings of 1975 was proof of the problem.

Quebec remained unconvinced about the need for the increases. Ontario also rejected Bell's evidence that the reductions in the construction program were necessary. In contrast to Quebec, Ontario suggested a $25.3 million cut in the proposed rate increases, while the Consumers' Association urged a total rejection of the rate increases. Consumer and interest groups were also opposed to the increases, using the opportunity of the hearing to remind the commission of its "duty to do much more than ... trying to help Bell each year to complete its Construction programme." With the formality and "ritualistic atmosphere" of the CTC hearings, participation was limited, and Bell could get away with not fulfilling its social responsibility.

The autumn 1975 hearings were conducted in an atmosphere of tension and resentment. Bell was thoroughly disgruntled at its treatment, while intervenors who opposed the rate application were frustrated at their seemingly consistent lack of success in getting the commission to take their views into account. The CTC itself must have known that whatever its decision, it would not satisfy anybody. In the end, the CTC approved the full amount, which represented a 6.5 per cent increase in the company's schedule of rates. The commission noted that the construction cutbacks were a "sideshow" that should be forgotten, but were irritating to the commission nonetheless. Geoffrey Stevens remarked that the CTC "appear[ed] to have been cowed" by the telephone company into capitulating to its demands.[68]

CONCLUSION

This was how the CTC ended its supervision of federal telecommunications regulation. By the time the CTC handed over its telecommunications responsibilities to the Canadian Radio and Television Commission in 1976, circumstances had forced the commission to deal with the consequences of several upheavals. The triangular contest between regulator, monopolist, and consumer meant that the maintenance of legitimacy and mediation could be accomplished only with great difficulty.

The ensuing conflict threw off the precarious balance between the economic and political markets. Bell's corporate objectives featured an overriding concern for procuring large infusions of new capital to finance expansion and modernization. The ethos of growth conflicted with demands that regulation hold the line on rate increases, distribute the burden between consumer and monopolist equitably, or direct telecoms towards socio-economic and political objectives such as rural development or support for low-income subscribers. Intervenors were agreed that greater public participation and accountability for regulatory outcomes would restore responsiveness to an atrophied system. There was markedly less agreement over the direction of technological development, or exactly how far state action should extend in remedying telecommunications regulation and its discontents.

The intense focus on the economic and political problems of telecommunications regulation also attracted a discussion about the moral economy of telecommunications. Frequent rate hearings provided consumer and special interest groups the opportunity to contest prevailing convictions about technological progress, price, and the pattern of diffusion of telecommunications from their perspective. The tide of consumerism, the articulation of demands from low-income groups, and the demands of provincial participants all shaped the tenor and contours of that debate.

On the other side, proponents of Bell's approach pointed to the success of the telecommunications industry in Canada. Here was a major component of the Canadian economy, owned by Canadians, and able to boast of significant successes in terms of diffusion of telephones, modernization, and at a reasonable cost.

The precise effects of the discussions on the proper social role of telecommunications in economic development are uncertain, but they seem to have had only a minor effect on regulatory results. They did, however, cast a long shadow over both Bell and the commission. Whether deservedly or not, company and commission suffered under public scrutiny. Consumers were likely to view Bell as less the provider of reliable telecommunications services and more the high-handed monopolist making excess profits by squeezing the subscriber. Public opinion of the CTC was not much different. The public was likely to view the commission less as the protector of the public interest, and more as a tribunal too preoccupied with protecting corporate profits. The image of both agency and utility inevitably suffered in the process.

By the time the regulatory crisis had reached the cabinet level in 1973, the system was on the verge of collapse. Attempted political solutions alleviated the situation, but only temporarily. Political intervention demonstrated a complete lack of confidence in the CTC's ability to regulate telecommunications. Although not revealed at the time, plans for immediate regulatory transfer to the CRTC illustrated the lengths the cabinet was prepared to travel to provide relief. The more modest measures eventually agreed upon were only partial indication of the seriousness of the problem.

CRTC regulation offered at least the hope that the transfer would mean change. The economic and political challenges that generated so much conflict transcended the boundaries of regulation; they would require much more than a remote hope of better things to come. Once lost, the symmetry between economic demands and the exigencies of the political marketplace would prove much more difficult to re-establish, and make the task of formulating a national telecommunications policy more difficult.

6 The Politics of Technological Development: Canada, 1970–1975

Telecommunications policy may have to be reshaped if full advantage is to be taken of the opportunities that technology affords and if socially undesirable effects are to be avoided. For, in the words of Francis Bacon, "he that will not apply new remedies must expect new evil; for time is the greatest innovator."

– Introduction to *Instant World:*
A Report On Telecommunications in Canada, 1971

We ... are aware that the many techniques of cybernetics, by transforming our control over data and information, may transform our whole society. With this knowledge we are wide awake, alert, capable of action: no longer are we blind, inert pawns of fate. A small thing, perhaps, but one which signifies a turning point in human development so important that it provides a foundation of new hope.

– P.E. Trudeau, "Technology, the Individual and the Party"
in Linden, *Living in the Seventies,* 1970

In an important sense, these two declarations convey both challenge and response. For the authors of *Instant World,* the key lay in adjusting Canadian telecommunications policy to changing technological reality. The state would have to channel economy and society into new realms of sophistication while restraining the possible side-effects of technological acceleration. Pierre Trudeau's response was both elegant and comforting: the state's very awareness of the dazzling possibilities of technological transformation would lead inevitably to action. No longer content to be "blind, inert pawns of fate," the federal state would emerge from the shadows to assert its role in shaping the technological future. At the same time, the specific course of action was non-deterministic; Trudeau promised only vigilance.

The technological pressures of the 1960s had resulted in three federal responses: a series of international agreements, a crown corporation dedicated to launching a domestic communications satellite, and a government department. In the complex process of defining Canadian and federal interests in microwaves and satellites, questions of control, use, and ownership became key issues. No less tangible were the new ambitions of policy-makers in the field. Those new institutions represented the footings of a comprehensive national telecommunications policy.

As we have seen, the Canadian postwar experience with technology had been marked by a certain ambivalence. Hopes of deliverance were usually tempered by foreboding about a future bereft of technological sovereignty, where control would be exercised by others. Understanding technology would provide the corrective; the challenge would be to incorporate that knowledge into a complete renovation of telecommunications policy. Given the right direction, federal telecommunications policy might guide Canadian business to exploit opportunities not yet realized, lead to a political renaissance of instant tele-referenda and decentralized government, and strengthen the bonds of nationhood. People could pour their hopes and fears into the vessel of technology. The promise contained therein knew few political or economic bounds.

This vision of a technocratic solution to the challenge of the communications revolution[1] was particularly appealing to policy elites, since they would be called upon to usher in some potentially momentous changes. The dizzying sensation of accelerated technological change and the sense of urgency it conferred offered a critical opportunity to unravel some of the more intractable political and economic problems in Canadian politics. Trudeau's argument, however urbane and reassuring, would compel the state to embark on an uncertain course of intergovernmental negotiation requiring a significant mobilization of the state's political resources. This chapter will explore the efforts of the federal state to translate those ambitions into policy in the area of telecommunications, focusing in particular on the traditional telecommunications common carriers between 1970 and 1975.

The impetus for federal reform came from two directions: the recognition that technological changes would require an organized state response; and a more generic but related desire to modernize and systematize the bureaucratic structure of the state in order to make it more effective. Its most articulate proponents were to be found in the French-Canadian political class, confident that the enhanced administrative capacities of the state could support regional or even national ambitions. When common convictions about the uses of the state were exercised for opposing ends, for example in Ontario and Quebec, conflict was inevitable. The result would be two states warring in the breast of a single technology.

Federal plans were soon challenged by provincial aspirations, which would become a prominent feature in attempts to construct a national telecommunications policy. Provincial politicians, especially in Quebec, were wary that such a policy might mean surrendering their own aspirations in a field perceived to be crucial to both their economic and cultural development. Dissatisfied with the preponderant federal role in telecommunications in its own province, Quebec repeatedly led the calls for the provincialization of control of all telecommunications as a matter of cultural security. The wave that prompted federal action on specific technological questions would not wash over regional interests and visions, nor extend to providing a mandate for the installation of a national policy in telecommunications. The smoke from the bigger battle over federalism dimmed chances of agreement over communications.

In the meantime, the telecommunications utilities such as Bell Canada, whose operations might be affected most dramatically by any changes, manoeuvred adroitly to forestall or delay any radical change. Their strategy lay in quiet lobbying of both federal and provincial governments to ensure that any policy would have the least possible impact on their activities. Such companies could also rely on federal-provincial inertia to delay moves which would alter their position in the marketplace.

If the technology of mid-twentieth century Canada was generated mainly under the influence of American innovation applied by large Canadian utilities, its politics were patterned primarily on the decentralized nature of its federation. By 1975, conflicting conceptions of federalism sundered any hope of federal-provincial agreement over a national telecommunications policy.

STUDYING TELECOMS

Although the question of the domestic communications satellite had been resolved by 1970, a more comprehensive structure was needed to cover the bewildering range of issues connected with telecommunications technology. To that end, the Department of Communications presided over a wide-ranging study of telecommunications issues, prospects, and implications. This section will examine the outcome of those studies, and the state's initial attempts to translate those discussions into policy.

The federal government's determination to focus on telecommunications led to the establishment of the Telecommunications Study Mission. Created by cabinet order in 1969 and nicknamed the "Telecommission," the study mission was chaired by the department's deputy minister, Allan Gotlieb, and Henry Hindley. The Telecommission embarked upon an examination of telecommunications that was both impressive and comprehensive. The goal was to define the problems the federal government would face in forging a new communications policy. The commission enlisted the services of a wide variety of public utility experts, communications specialists, industry representatives, academics, and policy-makers to write the forty-three studies on every possible aspect of telecommunications policy. Although the general report, entitled *Instant World: A Report on Telecommunications in Canada*, was at least partially intended for a wider audience, the document is instructive for the concerns raised about the importance of a restructuring of telecommunications policy.

The studies gave added justification to the creation of the Department of Communications. The new minister, Eric Kierans, stated that the DOC's creation was a recognition that communications had moved to the forefront of our national affairs and was now "the nerve system of a nation in search of itself ... [I]t is to develop that nerve system, to tune it, to make it work, that the Department of Communications is being formed." There was little question about the ideal of technological progress: the "promise of technology" required that we "have the courage to accept new facts" about the order of the technological world. "Technological progress cannot be retarded and still less can it be halted; it must be developed to the maximum, but it must be controlled." Kierans even envisioned technology "widen[ing] the opportunities for individuals to participate in the processes of government – by computer-calculated referendums or by information hot-line phones." Kierans warned, however, that "it is also likely that as government decisions become more and more dependent upon technique and specialized knowledge, so will genuine participation become impossible for all but a handful of experts. The difference between what may and what can happen will be decided by political will." The Telecommission could not possibly provide that will, but the resulting studies could provide the basis for far-reaching decisions to be taken in telecommunications policy.[2]

The commission did much to define the horizon of telecommunications in Canada. Deep thinkers leapt ahead to consider the possibilities of the next stage of technological development. On the other side, telephone companies such as Bell and Saskatchewan Telecommunications were much more reluctant to embark on radical departures, and exerted their influence to ensure that any changes would take into account the interests of the established common carriers.

The reaction to the Telecommission studies was on the whole positive. But at the first sign of deviation from the acceptable canon of policy options, groups such as the Trans-Canada Telephone System were quick to make their displeasure known. Communications theorist Dallas Smythe, a professor from Saskatchewan, had undertaken the writing of the Telecommission's Study 1(e) on Canadian telecommunications regulation. Smythe had advocated a significant expansion of the regulatory and planning role of state in telecommunications. Industry response was swift. "It is the opinion of TCTS and its members," Trans-Canada Telephone System representative C.E. McGee wrote the Department of Communications, "that Mr Smythe's recommendations for Canada are neither relevant nor practical when considered in the light of the realities of Canada's federal system and the telecommunications experience in this country."[3] If reform was inevitable, then at least the degree and quality of that reform could be controlled.

Bell in particular was careful to stake out its position, and defend its planning record in the telecommunications field. This was not surprising, since Bell had the most experience in the matter, and the most to lose. As G.E. Inns of Bell complained to Richard Gwyn, "[g]overnment officials have made public pronouncements insinuating that Bell and the industry in general has done no planning and that the Department is going to do the planning in the future. Such statements can only cause ill feeling and engender distrust and lack of cooperation." Bell's hope was to avoid dual regulation and maintain the status quo with federal and regional authorities.[4] The statements belied an increasing unease among the common carriers that the DOC was poised to intervene in the industry in spite of reassurances to the contrary.

The Telecommission studies helped to define the contemporary issues in telecommunications policy. A good number of the studies were technical in nature, but some established that, whatever form government regulation was going to take, the state would not engage in a nationalization of the telecommunications system on the model of European Posts, Telegraphs and Telecommunications (PT&TS). The commission established that the regulatory apparatus should more accurately reflect the principle of access, especially in the domain of new technologies. What this meant to Kierans was "that a continuing effort must be maintained to guarantee that the power of the individual, the ordinary consumer or the ordinary citizen is not diminished or constrained by the efficiency of machines serving the systems or the established social or economic order."

Perhaps the most important principle enunciated by the studies was the critical importance of Canadian control. The telecommunications industry was owned and operated by Canadian firms, its research and development establishment second to none. That unusual and happy state of affairs generated options for industry and government alike that could safeguard Canadian control and secure a bright future for telecommunications. Here was an industry with which Canadian economic destinies could be fashioned. Kierans and his successors in the DOC would attempt to link national interests with the promotion of the public utilities to produce an accessible and economically efficient communications infrastructure. The federal state had a special responsibility to do so, as Kierans explained, because of the strategic importance of communications. This was also a principle over which most could agree; indeed, telecommunications was one of the few sectors where almost total Canadian control was exercised. The dilemma lay in what action the state should implement to ensure control remained in Canadian hands.[5]

In the wake of the publication of the Telecommission studies in the spring of 1971, the federal government moved quickly to establish parameters for a national telecommunications policy. Department officials were aiming to introduce legislation in the 1971–2 session of Parliament. The DOC's objectives were sweeping, and included a broad outline of the principal features of a telecommunications act which would amend and consolidate existing federal legislation in order to "establish a single coherent body of telecommunications law."

In its preliminary design, telecommunications regulation was to be expanded to include interprovincial tolls. The Trans-Canada Telephone System had escaped any regulatory oversight since its inception in 1931; such a condition did not "afford proper opportunity for expression of the public interest." Nor did the TCTS promote the orderly development of telecommunications systems in Canada – at least not along the lines the federal policy-makers desired. The department acknowledged that a "commendable degree of coordination and standardization" had been achieved by the TCTS, but equally stressed that the time had come for a more rigorous assertion of the values and priorities of the state.

The policy document also expressed dissatisfaction with the continued CTC regulation over telecommunications. The arrangement resulted in a divided federal jurisdiction over communications and broadcasting, which were held to be much closer than the ties between transport and communications. Industry structure in transportation and telecommunications were also incompatible, and the expertise required for each was quite different.

Two alternatives were presented: either the creation of a Canadian Communications Commission to deal with telecommunications; or a reconstitution the Canadian Radio-Television Commission. The first alternative would still result in divided federal regulatory responsibility for communications. It would also likely foster possible conflicts between two independent and aggressive regulatory bodies. The preferred alternative lay in a revamped CRTC. Regulatory transfer would end the division of federal regulatory authority, and facilitate "a broader perspective than under present arrangements." Transfer to the CRTC would mean unified and more coordinated regulation, and act as a more effective tool of oversight in the field.

The possibility of the transfer of the CTC's regulatory powers over telecommunications had received attention for some time. The CTC chairman, J.W. Pickersgill, suggested the transfer in 1968, and Gérard Pélletier made the same suggestion while interim minister of communications after the resignation of Eric Kierans in April 1971. Technological developments had weakened the traditional telecom-transport link in favour of a closer relationship between telecommunications and broadcasting. That reason alone, Prime Minister Trudeau told Pélletier, "motiverait à elle seule une nouvelle répartition des pouvoirs reglementaires au sein des télécommunications."[6]

Trudeau cautioned that provincial reaction to any new communications agency might provoke jurisdictional disputes. Although consolidation of regulatory responsibility would not involve additional federal powers, nonetheless, the exercise of the regulatory powers might incite "ressentiment et opposition de la part de certaines provinces." While advocating forbearance in the approach to the issue, the prime minister wholeheartedly approved of the thinking behind the transfer. To policy-makers of the day, the fusion of culture and communications seemed natural. As early as 1969, Gérard Pélletier had spoken about culture becoming the "principal agent de mediation entre les Canadiens." Recent technological discoveries would add to the pedagogical and cultural revolution of the time. The challenge was to harness the new methods of communications that were penetrating all areas of human activity.[7]

The growing interaction between broadcasting and other forms of telecommunications persuaded the government to transfer ministerial responsibility for the CRTC to the minister of communications from the secretary of state. The move was a modest first step, and gave some indication of the direction of the federal government in telecommunications policy. In essence, policy responsibility followed Pélletier from his post as secretary of state to his new permanent position as minister of communications in late 1972. The move would centralize all communications policy functions in the DOC.

Any potential transfer from the CTC to the CRTC would also have to overcome provincial animosity towards the broadcast regulator.

The CRTC was aggressive in the execution of its mandate, and rigorously centralist in the preservation of its power. While the Department of Communications seemed to allow for some flexibility over the issue of provincial input into CRTC decisions, the CRTC chairman, Pierre Juneau, rejected any such attempt. "The CRTC position," one official wrote, "is essentially to retain the *status quo*." The regulatory agency assumed a vigorous national posture when dealing with responsibilities falling under its jurisdiction.

Federal ambitions over telecommunications policy began to take shape in this way. The first tentative steps involved the transformation of a series of policy discussions and studies into a coherent plan for action in the sector. In order for federal ambitions for a national telecommunications policy to be truly realized, however, potentially difficult negotiations with the provinces would have to be initiated.

COMMUNICATIONS AND PROVINCIAL ASPIRATIONS

Canadian political culture and a decentralized federation would put the Trudeau government's telecommunications policy ambitions to that uniquely Canadian endurance test of federal-provincial discussion and compromise. Not surprisingly, the first federal initiatives in communications policy between 1971 and 1973 drew a variety of responses from the provinces. Each saw the field as key to the development of regional and even cultural development aims. The communications revolution thus confronted the division of powers in a culturally and regionally divided country. This section will focus on both Ontario and Quebec reactions; but nowhere would this confrontation be more manifest than in Quebec.

Communications policy in Quebec in this period was developing in tandem with two things: *étatisme* and the keen wish to accelerate economic and social development in the province. As Kenneth McRoberts and Dale Posgate note, the "state was to be the *moteur principal* of this rattrapage." The Quebec state sought to assume the powers needed to implement its social and economic development strategy. Political modernization meant that the federal system would have to relinquish more authority. All of this affected the province's communications policy.[8]

Although the Ministère des Communications du Québec (MCQ) was created in 1969 after the federal department, the new Quebec department experienced its real efflorescence after the election of the Bourassa government in 1970. Although the Bourassa Liberals rejected the more *dirigiste* forms of *étatisme* as a philosophy of government, they nevertheless stressed the necessity for provincial oversight in such key economic and cultural sectors as communications. In matters of culture, Quebec Liberals were attempting to "ride the tiger": in other words, they were committed to the federal system but simultaneously sought to associate themselves with indépendantiste ideals by developing the idea of cultural sovereignty. Practically, this meant complete provincial control over culture and communications. Jean-Paul L'Allier, widely regarded as the most nationalist of Bourassa's cabinet members, was given the Communications portfolio, a position he held until the mid-1970s.

By May of 1971, the Quebec government had circulated a working document entitled "The Case for a Quebec Policy on Communications," which detailed the province's intentions. The tenor of the paper was almost apocalyptic, warning that if the province did not join the communications revolution, "it would be relegated to the rank of a modern tribal society, the unenviable lot of all societies that turn away or are prevented from following the path of modern progress."[9] An energetic communications policy would ensure that Québécois could "be themselves in their own province."

L'Allier's case for a communications policy was sweeping and total, encompassing the economic, cultural, and social priorities of the province. The MCQ grew in tandem with the modernization of the bureaucratic state as well. In 1969 the department did not even exist; by 1972 almost three hundred people were working for it, a fact of some considerable pride to both minister and department as proof of a rapidly developing technocracy, ready to assume major responsibility for telecommunications development.

In terms of telecommunications regulation, Quebec argued that the province should properly retain supervisory control over all telephone companies operating in the province, specifically under the Régie des Services Publics (RSP). The Régie regulated over 450 small and medium-sized independent telephone companies in the province, but amounting to a small percentage of total telephones. In preparation, the MCQ began to prepare legislative amendments to the Quebec Public Service Board Act to strengthen its communications focus, and "devote its energies to translating the demands of the ongoing social development into terms of public interest." Perhaps the most significant claim, and the one which would cause the most conflict, was Quebec's conviction that not only Quebec's legislative framework in telecommunications, but also the federal constitution should be changed. The province had to realize its goal of becoming "both an agent of change and a tool for governing relations" in Quebec society. In the communications field, that meant full control over cable television, as well as broadcasting and telephony.

L'Allier's claims were reinforced by political scientist Léon Dion, who predicted that the 1970s would bring two revolutions – "one in telecommunications, using satellites … and the other in information, brought about by the fantastic development of a new science: data processing." L'Allier was convinced that Quebec had no choice but to assert control over the realm of communications. The dimension that concerns us here is the telecommunications aspects, and this was needed by Quebec to "encourage and speed up the industrial and technological development of Québec in the area of communications and telecommunications equipment, …[and] to promote co-operation between Québec and other countries, especially France."[10]

In this vision, telecommunications regulation would be transferred from Ottawa to Quebec City and placed under jurisdiction of the RSP. Cable was a matter of contention: Quebec wanted to exert control, noting that cable systems would be considered a public utility. Putting the Quebec operations of Bell Canada under the Régie des Services Publics was considered desirable, but not yet a priority. When asked in February of 1971 about the portfolio, Premier Bourassa did not deny that communications issues were "convenient": the field had broad appeal and in the event that Quebec did acquire jurisdiction, the responsibilities would not be financially burdensome.[11]

The Quebec program was the product of the combination of a technocratic ethos and a nationalist one. Yet, ironically, the same assumptions that motivated federal policy motivated Quebec policy: that communications had profound implications for the survival of nations and peoples. The state was especially necessary in the case of Quebec to ensure technology and the francophonie worked in tandem. The policy was both an economic and cultural imperative.

As part of Quebec's strategy for control of communications, L'Allier announced in mid-1971 that the government would propose three communications bills aimed at all aspects of communications in Quebec. Bill 37 was designed to give the RSP control of all communications in Quebec as well as putting control for broader communications policy in the hands of the provincial cabinet.

After the formation of the federal Department of Communications, Ontario also began to develop an interest in communications policy. The Ontario Telecommunications Committee was formed in 1969 to examine questions of communications policy. The committee recognized that there was a very important provincial interest in telecommunications and that the provinces should therefore develop and elaborate a provincial policy. The committee complained that the formal federal-provincial relationship in the area of telecommunications was limited to hearings before two federal regulatory bodies, the CTC and the CRTC. "More often than not," the committee report to cabinet stated, "this situation has resulted in provincial interests being ignored in the development and implementation of telecommunications policies." The committee identified regulation of common carriers, the establishment of the wired city, and the interconnection of various telecommunications systems as meriting the attention of the Ontario government.[12]

The push to define Ontario's interests in a field with such cachet generated a broader enthusiasm that infected even Ontario's bureaucrats. "One might declare," one Communications official later told a crowd of Ontario public servants, "that the decision to establish a department dealing with communications by the government of Ontario was an eager reaching out to embrace what some call the 'post industrial revolution.'"[13]

The decision to pursue the field more energetically had been developing for some years and was traceable to a discussion surrounding the advent of educational television in the 1960s. This later led to the interest in the "wired city" concept of multilinks between cable television, data processing and computer utility networks. A corollary objective would be to make sure the northern part of the province was not neglected.

The federal government's establishment of the Telecommission and its attendant set of recommendations and consultations prompted Ontario to awaken its interest over problems of communications.[14] A great deal of jurisdiction still resided with the provinces, and although Ontario's telecommunications infrastructure was predominantly managed by the federally chartered Bell Canada, the Ontario Telephone Service Commission had jurisdiction over many small telephone companies, just like the RSP in Quebec. The Ontario government also controlled a sizeable portion of telecommunications systems in Northern Ontario through the Ontario Northland Transportation Commission. But the federal government's energetic promotion of communications reform finally alerted provincial policy-makers that the time had come for more active provincial involvement.

This concern with developing a telecommunications policy for Ontario was enunciated in the speech from the throne in March of 1971. With the formation of the provincial Ministry of Transportation and Communications (MTC) in 1971, the government of William Davis signalled that Ontario would seek to protect its interests and develop a communications policy for Ontario which would "ensure that the interests of the people of Ontario are fully represented." Davis noted that his government needed to assess the real needs of Ontarians and to contrast those needs with the reality of telecommunications under federal jurisdiction. Transportation and Communications were fused together, with the rationale that both dealt with the movement of people and information and both needed similar regulatory tools.[15]

By 1972 the objectives of an Ontario policy in telecommunications were taking shape. Apart from ensuring the maximization of services, the Ontario government linked telecommunications to "strengthen[ing] a sense of provincial and regional identity by informing people of issues and problems." The other objectives were as modest, including the maintenance of just and reasonable rates in telecommunications, ensuring Canadian control of communications services, and orderly planning. The government also desired the system to be more responsive to economic development goals, and to new technologies and services. These objectives also stressed that Ontario wanted to be consulted in the regulatory arena when matters affected the provincial interests.[16]

Ontario's position as intervenor at rate hearings was no longer enough, especially since intervention put the province in an adversarial position instead of in the role of constructive policy-making. Clearly, the jurisdiction of the CTC was preventing Ontario from responding to its objectives. Although not contesting federal authority, Premier Davis noted that his government could "also appreciate that exploding technology [had] sufficiently surpassed the technicalities of the Constitution and related jurisprudence to render doubtful in some areas the unqualified right of federal regulation alone."

The areas of interest to the government of Ontario had a familiar ring to them. The province was interested in regulating telecommunications as a tool of social and economic development. It was also interested in enlarging its role in "establishing and/or regulating" computer utilities. Still, the government was slow to define a communications policy, at least compared to its Quebec or federal counterparts. Gordon Carton, McNaughton's successor at MTC, could ask:

Why should the province be involved, especially when we in Ontario – we in southern Ontario at least – are among the best served people in the world? Our communications system virtually places the entire world at our door. We have the widest choice and the finest quality of information systems available to use. Up to now, this had occurred substantially without the involvement of the province.[17]

Ontario's responsibility for economic, social, and cultural planning required the province to have control over the "machinery to achieve provincial goals and objectives." Of course, this did not mean serious revision of Ontario's policy of emphasis on markets to provide the essentials. Events had occurred and would continue to occur without the substantial involvement of the province. But clearly provincial objectives were to be more clearly defined and applied in the communications field.

The main preoccupation, voiced time and time again, was to ensure that the province avoided duplication and safeguarded the health of the telecommunications industry in Ontario. The basic goal of Ontario's telecommunications policy was to demonstrate that there was a material public interest involved; that "reasoned and reasonable" policy goals and objectives be adopted; and that workable and satisfactory intergovernmental arrangements could be approved.

Ontario's short-term posture, then, was one of a vigorous watchdog of the province's interest. But this did not include pushing for jurisdiction over telecommunications. Instead, the province contemplated urging the federal government to exercise its own responsibilities, while Ontario would exercise the more modest role in the telecommunications system via its control over Ontario Northland and the small independent telephone companies. Ontario's approach to the federal government in communications matters was limited to obtaining a better balance of power between governments. Implementation of this strategy was hampered by the fact that the province did not regulate the preponderant portion of the industry. Instead, the province's influence over the industry came "simply because we are the Government of Ontario," and as such was accorded a certain automatic influence. At the federal level, the province sought to negotiate to secure provincial legislative control over intra-provincial communications such as cable, and attempt to ensure that new federal legislation would incorporate provincial views.

Despite Ontario's position over the importance of controlling telecommunications policy, the province was not unhappy with its limited power over telecommunications. Despite several grievances over specific issues and a genuine desire to see serious federal regulatory reform proceed, Ontario officials were ultimately indifferent over questions of jurisdiction and control. So long as Ontarians enjoyed a high level of telecommunications services, their provincial authorities were happy to exert pressure from within, and resist calls for a root-and-branch reform of the entire national telecommunications system advocated by some of its provincial counterparts.

SHAPING GOVERNMENT POLICY

By early 1973, the federal government had completed its first round of consultations, and was busy preparing a major communications policy paper. For the moment, however, Pélletier stressed that the short-term intentions of the government would be directed towards the improvement of the regulatory measures which directly affected telecommunications.[18] It was certainly a good place to start.

After numerous interdepartmental meetings and outside consultations, a paper entitled "Communications and the Future of Canada" made the final rounds among interested departments and agencies. The draft document, two years in the making, stressed the impending arrival of technological developments that would add important new dimensions to the concept of telecommunications. The changes had the potential of being "deployed to best advantage" so long as a coherent national policy were in place. The foundations of such a policy would be provision of diverse sources of information; the development and preservation of high-quality telecommunications systems; the economic deployment of available human and material resources; and the assurance of a Canadian-controlled system. This, the paper argued, was the cardinal reason that technical and economic aspects of communications could not be treated separately from their social and cultural implications. The problem would be to decide which set of priorities would take precedence. Jules Léger, the under-secretary of state, argued strongly for the primacy of social and cultural values over econo-technical aspects, where conflicts existed. Communications officials, however, balked at giving such an absolute commitment.[19]

The document dealt at length with the implications of broadcasting and cable innovations, as well as the importance of retaining Canadian control. It lamented the fact that the recognition of the national dimension had been largely left to the Trans-Canada Telephone System, and recommended a greater regulatory role. Any new agency retaining responsibility would have to ensure effective economic regulation, especially of new services. The draft paper also tackled the issue of the interconnection of private systems, new types of equipment, and "foreign" attachments to telephone lines. The conclusion was that some sort of liberalization was desirable, but the government only went halfway by proposing that the regulatory body assure that interconnection complied with technical standards, and be in the public interest.

On the subject of a single federal agency for communications, the government envisioned a need for a more rigorous form of "continuing surveillance' of the performance of carriers. Since the link between communications and transport had already been severed in the government's executive arrangements, the paper restated the department's commitment to regulatory transfer. The move would also clearly contemplate the possibility of a single agency, taking into account of the increasing interaction between broadcasting and other forms of telecommunications, especially cable systems.

Significantly, the document also sounded the government's willingness to consider the delegation of federal authority over purely intra-provincial matters to an appropriate provincial regulatory body. In order to possibly forestall greater regulation, the Trans-Canada Telephone System representatives suggested the formation of a National Association of Telecommunications Regulatory Authorities (NATRA), which would have consisted of "delegates from federal and provincial regulatory bodies, to consult on matters of rates and service, and on national and provincial objectives, taking regional differences and requirements into account." The suggestions were incorporated into the federal proposals.

Finally, the paper made mention of the time that had passed since the intention to do something about telecommunications in 1970–1. Since then, the paper argued, the government had moved from committing itself to detailed elaboration to outlining the principle the government intended to follow in a national communications policy. The appeal ended on a somewhat provocative note, saying that "while the fullest possible regard must be paid to the special needs of different parts of Canada it is the national interest that must prevail." The DOC used the phrase intentionally, arguing that it "may be a good thing to end a Paper which is positively oozing with goodwill to the provinces and might otherwise be regarded as a measure of 'appeasement'."

The draft proposals generated some concern, especially among policy-makers charged with defending cultural policy. Jules Léger was most concerned that a clear commitment by the government "to the primacy of the cultural and other social values" was essential. Any possibility of a perception of a downgrading of cultural matters clearly worried the secretary of state. Pierre Juneau also had the overall impression that "concern with present and future cultural conditions are still not reflected strongly enough in the logic of the paper." Juneau's perspective was predominantly concerned with broadcasting technologies; nonetheless, the problem would have implications for the entire field. "We must realize," Juneau wrote to Allan Gotlieb, "that we are delegating to one increasingly unified field, tremendous cultural power." The result might alter the prevailing technologies and cultural arrangements "beyond recognition." Juneau's principal concern was that the logic that in the past had guided communication innovations along technical and economic lines was excessively beholden to the marketplace. Other goals – such as the need for domestic ownership and control, as well as national unity – had to assume a new prominence. The CRTC chairman pointed to the transportation field and how the "lack of social and cultural indicators" in the formulation of transport policy had led to an overemphasis on developing infrastructure for automobiles at the expense of public transportation. Juneau relied upon the writings of Harold Innis to make the argument that cultural orientation would undoubtedly be affected by the "physical aspects of message origination transmission and reception."[20]

The federal government's green paper was finally released in March 1973, and contained most of the recommendations previously circulated. The document stressed the importance of the maintenance of east-west links, including Canadian ownership and control of the country's telecommunications systems. The government was also at pains to stress that national objectives in the cultural and social realms would of necessity require a concerted federal response. Above all, protecting the national interest would naturally require the assertion of an explicit national dimension in telecommunications policy. Despite this, the green paper expressed the wish that questions of legislative and regulatory authority be avoided, since in any possible dispute, "the real ultimate loser would be the Canadian public." The paper also affirmed the state's intentions to proceed with internal regulatory reform, by combining the telecommunications functions of the CTC with the CRTC so that one regulator would be in a position "to take account of the increasing interaction between broadcasting and other forms of telecommunications." As discussed in the draft document, two-tier regulation was offered as an alternative as well, but was up for discussion.[21]

Reaction to the green paper was mixed. In the Standing Committee on Transport and Communications, the minister of communications had to answer charges that provincial momentum over telecommunications issues had increased because of the existence of a "vacuum of policy and/or definition of policy" on the part of the federal government. The charges were particularly pointed in view of the fact over two years had elapsed since the intention to pronounce a policy on communications. Pélletier defended the timing of the proposals, especially since the DOC had to wait for provincial input, which was slow in coming.[22] The minister's explanation sounded reasonable, but there was undoubtedly more than a grain of truth to the charge that a policy vacuum had existed. Pélletier and his department had accepted the calculated risk (or the necessity) of attempting to obtain some measure of provincial consent before proceeding with their substantive proposals. On the other hand, the length of time that had elapsed was an early indication of a weakening of political will.

Pélletier nevertheless signalled a willingness to act, and soon, over telecommunications matters. Time was of the essence: developments in teleprocessing and electronics threatened national sovereignty, and would carry social and cultural impacts that could not be left unattended. In the past, the telephone had been considered as "standard household equipment more like steps leading to the sidewalk"; Pélletier promised that technology would be more consciously put to the service of the country.

The federal government's policy statement elicited a mixed response from the provinces. Almost immediately after its release, L'Allier was preparing to take a very hard line on the federal proposals. *Le Devoir* had already called the proposal a "declaration of war" upon Quebec, a measure of the trouble the federal proposals would encounter.

Ontario's reaction to Ottawa's proposals was also mixed, but a good deal more positive than that of Quebec's. "Generally speaking," MTC's W.A. Rathbun commented, "many, if not most, of the proposals would improve Ontario's position, if adopted." But Ontario foresaw that unanimous provincial consent would be difficult to achieve. For example, the two-tier system that was proposed (not enthusiastically) might have been acceptable to the two central Canadian provinces, but would result in a significant erosion of the power of provinces such as Saskatchewan, who ran their own telecommunications utilities. The alternative proposal, a consultative mechanism for federal and provincial telecommunications coordination, might be acceptable to prairie and maritime governments, but fall far short of Quebec and Ontario aspirations in the sector. The green paper might have been targeted to frustrate plans for a unified interprovincial front by offering some practical concessions in exchange for provincial consent.

In the wake of the federal initiative, DOC officials met with their provincial counterparts to try to maintain the momentum the release of the paper created. The phrase that most rankled provincial sensibilities was "national objectives." Ontario and Quebec were also concerned that the proposed transfer of telecommunications regulation from the CTC to the CRTC was not simply a matter of good housekeeping, but a fundamental shift which dramatically affected the provinces, especially since CRTC and CTC rules of procedure were so different.[23] Ontario officials became convinced that Ottawa was determined to arrive at some kind of new policy regardless of provincial objections. Gordon Carton, the minister of the Ontario Ministry of Transportation and Communications, insisted to his federal counterpart that the province really wanted more direct participation in the regulatory process. Two-tier regulation was on the table, but seemed a distant contingency.

The growing provincial concern over their own telecommunications interests required the full airing that a federal-provincial conference might bring. Accordingly a conference was scheduled for November 1973 in Ottawa. It would give the participants a public opportunity to advocate their respective positions and persuade public opinion to their cause. Some even hoped that the conference might yield tangible agreements over a series of pressing telecommunications issues.

Interprovincial strategy in preparation for the conference focused on attempts to present a common front against possible federal intransigence over jurisdictional issues. Quebec and Ontario worked closely together, but Quebec was the most serious about wresting jurisdictional control from the federal government. L'Allier was confident that Bell was ready to come under provincial regulation, an impression undoubtedly reinforced by the whisperings of Bell executives who tried to play one level of government off against the other. The move would have to be predicated upon Ontario-Quebec solidarity in pursuing their common objectives. L'Allier insisted that the two provinces could not be split in consultation and negotiation with the federal government, and especially in regard to jurisdiction over Bell. The Quebec minister insisted to his Ontario colleague Gordon Carton that his government was not interested in a showdown, but was willing to partake in "patient" discussions as long as explanations were forthcoming from the federal government.

L'Allier's own assessment of a possible united front of provinces was not encouraging. In his estimation, the Maritimes were a mess of semi-feudal dependencies in thrall to the federal government; Nova Scotia "lagged behind" the provincial position, and was markedly more pro-federal, while New Brunswick's need for federal subsidies muted any possible opposition from Moncton, especially over a low-priority issue. On the other hand, British Columbia, Alberta, and Saskatchewan might be counted on, but similar problems might dissolve any attempt to join in solidarity. Prairie communications ministers also exhibited a very cautious attitude towards federal initiatives; they were particularly suspicious of the federal motivation for pressing the interconnection issue. Most of all, perhaps, western ministers were concerned with the preservation of their telephone companies as the chosen instruments they were intended to be.[24]

Quebec's hardening position over jurisdiction was reinforced by Premier Bourassa's strategy of pursuing cultural sovereignty and security via devolution of federal power. To that end, L'Allier sought to obtain a type of special status for the province in telecommunications matters by suggesting that provinces not wishing to assume control over any telecommunications matter Quebec insisted was provincial could delegate the responsibilities back to the federal government. A two-tier structure was also advocated, with a heavy emphasis on the provincial component.

Ontario officials were clearly uneasy with Québec's position, which was destined to meet with federal denunciation. Ontario officials feared Quebec might stiffen the federal government's position, and the greatest loser would be Ontario. Further, the MTC concluded that the federal government would never be seen to have capitulated to provincial demands, especially ones that appeared to favour a special status for Quebec. Worst of all, the provinces would be outmanoeuvred if Ottawa decided to portray provincial attempts to win telecom jurisdiction as a blueprint for the balkanization of an important national resource. Ontario feared plans made in Quebec City would sink its chances at rationally discussing matters like jurisdiction over cable. Meetings with federal officials seemed to confirm that judgment.

Within a few short years, communications policy had moved to the top of the national agenda, but not the way Eric Kierans had envisaged in 1969. When Pierre Juneau wrote in Le Droit that "l'année 1973 sera une ligne de demarcation entre les communications anciennes et les nouvelles," he could easily have included that it would also be the demarcation line for the battle between Ottawa and the provincial capitals over communication, the first major public manifestation of which would be the federal-provincial conference on communications.

The conference opened in Ottawa on 29 November 1973. Pélletier began by stressing that the growth of the Canadian communications structure was at stake: "The principle of respect and concern for the growth of Canadian structures will in the end guide the Canadian communications system into the generation of scientific and economic development." He was careful to stress the national as opposed to the regional nature of the Canadian system. The conference threw into relief the difficulties that would plague successive attempts at reaching a comprehensive solution to the telecommunications problem. "We are administering telecommunications," one commentator observed, "with the tools used in the time of Morse and Marconi. And although they have been effective in the past, there is reason to doubt that they will stand up in the future."[25]

The provincial reaction to the federal proposals was predictable, but nonetheless disappointing for federal officials. Quebec and British Columbia led the provincial charge, arguing forcefully that provincial, not federal, control was more suited to the telecommunications field.

For his part, Pélletier indicated that the federal government was not prepared even to consider vacating the field. The irony was not difficult to discern: the green paper has commented that there was "nothing to be gained from the adoption of rigid attitudes leading to disputes about the distribution of legislative and regulatory authority from which the real ultimate loser would be the Canadian public." But in the space of seven months, the warning turned out to be prophetic.[26]

It was plain from 1972 onwards that a government with a mandate to fortify national unity and sovereignty would collide with provincial governments who had particular regional or cultural imperatives. Both Quebec and Ottawa believed that control over communications was crucial to their respective goals of cultural sovereignty and national integrity. The goals were different, but the assumptions about the technology were the same.

The conference was thus doomed from the start. The meeting pitted a federal government whose communications policy had high priority on its legislative agenda against provinces that had varying goals, but a common front. "Le statu quo sur cette question," wrote journalist André Beliveau, "nous est inacceptable. Si cette situation se mantient, nous risquerons, quant à nous de devenir un peu moins qu'un gouvernement, et un peu plus qu'un group de pression." That aptly summed up L'Allier's position. Ontario had put the question of jurisdiction at the bottom of its list of priorities, but even so, the province was in no mood to acquiesce. "The fact is," Gordon Carton argued, "that communications as an industry is so important as part of the social and economic fabric of our province that the government must be involved."[27] Saskatchewan's position on cable systems was also firm. National and regional objectives in culture, economic development, and political control would come into conflict.

When Pélletier had finished, the Prince Edward Island minister of communications, Paul Creaghan, read a joint provincial statement that expressed the desire for more provincial control and the hope that this could be achieved through a political settlement.[28] Ontario stressed its desire to control cable distribution systems and its dissatisfaction with the current regulatory regime in telecommunications. Quebec reiterated its demands to control the entire sphere of communications in the province.

Predictably, the conference ended in stalemate. Pélletier announced that "the scope asked for by Quebec and British Columbia, unless I'm mistaken, is to say get out of there completely so I can take your place. No we are not ready to consider that because we believe that there absolutely must be a federal presence if we wish this country to keep together." Loss of jurisdictional control for Ottawa would be a serious blow to its mandate to preserve national unity. Jurisdictional realignment, at the heart of Quebec's ambitions, would either wait for a convergence of the positions of the two governments, or a reference to the Supreme Court.[29]

Since the federal government was anxious to regain the initiative, federal-provincial meetings continued apace after the 1973 conference. In an attempt to shore up the foundering federal position, Ottawa initiated a series of bilateral meetings throughout the winter of 1974. The federal strategy was geared to steering provincial ambitions away from any constitutional reallocation of responsibilities; discussion had to begin with the status quo. Further, Pélletier and his officials sought extensive clarification of provincial positions. He also wanted to make clear to the provinces that the proposal for two-tier regulation, which the feds had not been enthusiastic about in the first place, was no longer an option, at least in the areas of telephone/common carrier regulation or broadcasting.

Ottawa's renewed communications offensive of 1974 was interpreted in provincial capitals as an attempt to turn federal-provincial discussions in to a series of "hard-nosed horse trading sessions" over concrete communications issues. Ottawa was now proposing another conference, this time *in camera*, of communications ministers. The publicity surrounding the 1973 conference had not served federal strategy well; moving communications problems from centre-stage might get things moving. From the provincial point of view, the federal proposals could emasculate important issues in interministerial committees "which could run on forever." In essence, the federal position in the spring of 1974 offered no substantive changes, and stonewalled any prospective provincial initiatives in the field.

Reaction to Ottawa's stiffening stance was predictable. For Ontario, the federal position had caused serious disappointment. The province was willing to discuss the possibility of a closed federal-provincial conference, but only if real progress was a possibility. Communications policy objectives by 1974 had expanded to include the promotion of a "better understanding between Ontarians, Canadians and with the global community," orderly planning, and maximum Canadian control of communications services.[30] The objectives were limited to ensuring a full participatory role for the province in both federal policy and the regulatory process where it touched on the provincial interest.

Ontario's concerns were also electoral in nature. The MTC minister John Rhodes, Carton's replacement, wrote Premier Davis in mid-1974 to advocate a strong position in federal-Ontario negotiations over communications. The minister strongly recommended that major legislation be introduced in the fall 1974 session of the legislature. Federal-provincial communications talks had reached an impasse.

Rhodes was mindful that a provincial election would have to be called in the succeeding twelve to eighteen months; Ontario's communications strategy would have to integrate itself with Davis's electoral designs. "Do we wish a public conflict with the federal government and a potentially major debate in the legislature on communications during the next year?" Rhodes asked the premier.[31] The minister warned of the communications industry's increasing unease with the lack of federal or provincial policies.

Quebec's position was much tougher. The province initially agreed to a closed conference, then reversed itself, probably after discussions with British Columbia. Both provinces maintained that the only way to bring pressure to bear on a government reluctant to deal was by keeping the issue alive in the public mind. *La Presse* commented that this fight was not just about communications, but the "very essence of confederation applied to the field of communications." L'Allier commented that to go any further "it would be necessary to leave the federal framework."[32]

Pressure on the federal government began to mount, especially after the Interprovincial Communications Ministers Conference in May 1974. Ontario's new-found willingness to promote a more forceful attitude was perhaps the most serious indication that positions had becoming increasingly rigid. "While this shift in emphasis does not move the communications sector to a position of paramountcy as has been the case for Quebec," one Ontario official observed, "it does represent a significant hardening in terms of provincial responsibilities in all aspects of communications." The shift in attitude was backed up by John Rhodes's declaration that the government of Ontario was willing to go it alone on communications if a united provincial front was not possible.[33]

Ontario's more aggressive posture was warmly welcomed by Quebec officials, and likely did much to promote provincial solidarity over communications. Ontario asserted the need for more provincial control over intraprovincial common carriers, but with continued federal involvement in the interprovincial or international spheres. British Columbia desired full control over BC Tel, and the prairie provinces reiterated their stout opposition to any federal intrusion in the operation of their provincial common carriers. Perhaps most frustrating for the provinces was the fact that little or no progress had been made even where there had been some promise of ready agreement. For Robert Strachan, British Columbia's minister of communications and spokesman for the interprovincial conference, the situation had developed out of Ottawa's failure to "examine and discuss seriously the proper role" and responsibilities of the provinces in the communications field.[34]

The practical outcome of the provincial round of meetings over communications was to demonstrate that, despite some movement, the provincial positions remained close on the questions such as cable jurisdiction but much farther apart on the issue of telecommunications. Provincial solidarity might easily be asserted when general principles were discussed, but this quickly disintegrated when specific issues were brought up. Ontario was the only province, despite its new posture, where a federal presence in the common carrier field was felt necessary and desirable. Quebec, in tandem with the prairie and maritime provinces, accepted no federal role.

The differences between provincial governments were accentuated by the different styles of regulation each province preferred. Ontario viewed prairie telephone administrations as "nothing more than monopolists extracting monopoly profits from the rest of Canada" which was definitely not "in the interests of Canada or Ontario." Québec viewed greater competition in telecommunications as costly and unnecessary, and according to Ontario officials, probably based on the "[incorrect] assumption that Quebec could regulate Bell effectively." Discord over the general direction of telecommunications regulation was bound to cause stresses in any provincially based strategy to combat a national telecommunications plan. One conspicuous example of provincial disagreement over specific telecommunications issues was the question of interconnection.

INTERCONNECTION

The problem of interconnection of systems provides a valuable opportunity to examine the play of conflicting interests between governments and industry on an important telecommunications problem. The triangular relationship of industry, provincial, and federal governments is perhaps plainer to see in discussions over interconnection than in the more straightforward federal-provincial talks. On this issue, the federal government seized the high ground, arguing for more open architecture. Conversely, provincial discord and disagreement slowed the liberalization of interconnection of non-telephone company owned attachments and systems.

Pressures for interconnection in the United States were partly a consequence of the acquired industrial communications capability that came with the Second World War. Industries developed markets to challenge the telephone industry's right to refuse to connect customer-provided equipment. The 1956 Hush-a-Phone Case before the US Court of Appeals resulted in a ruling that confirmed the right of the customer to connect equipment "privately beneficial without being publicly detrimental." The next American milestone came in the mid 1960s with the Carterfone decision. The Carterfone was a device used to link private mobile radio systems with the telephone network by acoustic or inductive coupling. The Federal Communications Commission ordered the Bell system to liberalize the terms and conditions for interconnection. The Canadian response was to liberalize some tariffs in 1969, but keep the system intact.[35] The Canadian courts, as early as the 1954 Sterl-a-Fone case before the Supreme Court, had protected the telephone company's right to refuse terminal attachments of any sort.

One Telecommission study on interconnection noted the rapid expansion of "mechanized information handling systems" which used remote access terminals and computers. The second factor the study identified was the plethora of new telecommunication terminals. Interconnection practices of recent vintage were creating problems, especially for the user's practice. By far the most unsettling development to the interconnection issue was the proliferation of data-processing installations.

Both user and manufacturer were often pitted against the telecommunications common carriers, who were more concerned with systems planning than with permitting free access. Carriers were careful to defend the possible "technical pollution" of the network from users which could result in a degradation of service. The necessary conclusion to this argument was that the only safe means of safeguarding the network was through carrier ownership of all terminals. The carriers would also factor the cost and maintenance of their equipment into the structure of local telephone service rates.

"The carriers' interconnection practices became a serious problem for the user," one Telecommission study noted, "when he discovered that he could not combine the optimum choice of communications terminal and carrier facility." Carrier practices relating to interconnection had "generated a triangular set of forces between the carrier, the user and manufacturer, which cannot stay in equilibrium without continuous attention and negotiations on the part of all three." The explosion of data communications, and the proliferation of private systems and potential common carriers, led to pressure upon the federal state to give general permission for interconnection. The study recommended that further studies be done, but with a view to encompassing the legitimate needs of the carrier, the manufacturer of telecommunications equipment, and the users. The report also advocated that standards be set for the interconnection of systems.

Bell's initial strategy was to argue that a general liberalization of interconnection would run counter to the public interest in both the technical and economic senses. The company's greatest fear was that a liberal interconnection policy would promote competition in lucrative markets while leaving the company with the statutory responsibility to service uneconomic areas of Bell territory. For Bell, interconnection represented the thin edge of the wedge in terms of competition. Approval would allow the connection of customer-provided equipment to telephone facilities. The federal state was under increasing pressure from suppliers and customers alike who desired greater access for customer-provided equipment. Bell regarded the pressures as a tangible result of the liberalization of interconnection policies in the United States. The company was particularly anxious to remind both the public and the government that a simple adoption of American policies would be a foolhardy venture.

Bell's policy by the mid-1970s was to allow limited interconnection, on its own terms. The policy would also have to "ensure the safety of employees and customers, and protect both our customers and the Company against negative economic effects and deterioration of service through electronic pollution and interference." The company dearly wished to avoid the American example of parvenu operations cream-skimming the more lucrative parts of the trade.

By mid-1973 the company was busily planning strategy for a possible change in the rules. "What are we now?" a Bell strategy document asked. "A little bit pregnant and close to being a 'Bastard!'" came the reply, since Bell was partly a regulated monopoly in telephone business, partly a regulated competitor in wire services, and partly a non-regulated competitor through Northern Electric and companies like TeleDirect. There were several alternatives open to the company: to strive to maintain the status quo; to foster a competitive environment; or to acquire a monopoly in facility provision, offering many competitive services. In that scenario, competition would exist in the provision of "added-value" businesses such as terminals, data services, consulting systems design, and the like. If competition were the outcome of the current discussions, Bell would insist that rates be deregulated, that its obligation to serve any market at "average" prices be removed, together with a full "recognition of the social and political consequences."[36] Bell's short-term strategy was simple: promote the continuation of the status quo, while implementing a "profit generating/strategic penetration" program which would include avoiding increases in message toll rates while raising local rates. "So far," the company concluded, "we have largely failed to achieve this objective."

The company's strategy was to also maintain extreme pressure on CN/CPT on digital service in order to weaken their bargaining positions for later negotiations. The same went for Telesat and Canadian Overseas Telecommunication Corporation by continuing to deny systems interconnection. This would mesh nicely with the possibility of achieving holding-company status for Bell without the overall regulation of the company's rate of return. This would effectively reduce the risk of a possible provincialization of telecommunications regulation. In the meantime, however, any pressures to interconnect Bell's system with CN/CPT or any other concern would be resisted with all the resources at the company's command.

Bell officials also contemplated the possibility of adopting a liberal posture on interconnection but at the same time making the consequences of such a policy – deregulation of toll rates, rate rebalancing between local and long distance rates, among others – well known. In that event, Bell would have to watch its credibility gap, especially since the company itself was not firmly convinced that interconnection would be the right move.

Industry reaction was generally favourable to some sort of liberalization of interconnection conditions. The telecommunications carriers, especially Bell, were on the whole much more reluctant to abandon their commanding position in the market. Why should they? Bell agreed that some liberalization should occur, as long as the carriers were not restricted in pricing and "also with the understanding that we are talking about terminal equipment only." Other associations such as the Electronic Industries Association of Canada wanted to ensure that the federal government understood that domestic industries should be given priority over international competition; as such they favoured the "principle of an orderly development of the interconnect market, favouring the employment of equipment made in Canada, in line with our 'Buy Canadian' Policy."[37]

Perhaps the most significant pressures for interconnection came from the computer/communications sector. Technological innovation had created a much starker line of demarcation between the common carriers and the providers of such services as time-sharing and data processing over telecommunications lines. Bell in particular came under sharp attack from the industry, especially over its restrictive policies over interconnection as well as delays in delivery of equipment. The problems were at least partially the outcome of the company's fear of cream-skimming. The proportion of Bell business devoted to data communications was relatively small, approximately

5 per cent; however, the sector was growing by about 70 per cent a year. Bell had established a Computer Communications Group to supervise the growth, but the problems of an increasingly dissatisfied supply market still remained.[38]

IBM Canada also favoured a liberalized interconnection policy, once the company could be assured that the integrity of carrier networks would be protected. A more liberal policy would undoubtedly be helpful in encouraging expansion and development of data processing applications and services, and hence a major benefit to IBM. The concerns of companies such as the Harding Corporation, manufacturers of telecommunications equipment, were not atypical. Harding complained that Bell had established a barricade of restrictive tariffs, principally achieved by "railroading the Canadian Transport Commission" into approving most non-rate oriented tariffs requirements by the Company. "Bell Canada's power," wrote an exasperated R.W. Walton, Harding's general manager, "has reached dictatorial levels and the inability of the CTC to properly police Bell's activities has allowed Bell to explode into the giant company it is today."[39]

The Canadian Industrial Communications Assembly weighed in with its opinion. Some CICA members complained that Bell was extremely reluctant to give away anything on the issue and that the company assumed that "interconnection is, at least, naughty." The telephone company's argument that its stand on interconnection flowed from its natural monopoly status was rejected out of hand. "The telephone company," J.M. Hillier argued, "was created through a) the entrepreneurial talent of those early businessmen who saw a change to make a buck out of the new telephone craze, and b) the obvious need for a single network connecting all phones to avoid duplication and simplify switching. It's the network that's monopolistic, not the instrumentation, or its supplier network."[40]

The federal Department of Communications recognized that the question of interconnection opened up a century-old practice that stressed a portal-to-portal monopoly which accented hardware ownership operation and maintenance. It was only in the 1960s that developed countries began accommodating separate ownership in the quasi-private terminal sector of carrier operations.

To the federal government, liberalization of interconnection would foster technological innovation. The question of innovation had been a prominent one at the inception of the Department of Communications. Greater access to the network might bring "refreshing continuing innovations, particularly where the attaching terminals are of a complex nature." Further, one aide memoire enthusiastically noted, the diffusion of sources of hardware supply would promote the revitalization of the telecommunications sector. Some forecasts indicated that by the 1980s up to half of the total assets engaged in telecommunications could be in the terminal area. "Any pronounced shift to the private sector," one DOC official noted, "would ease the burden of capital requirements from the carriers." Interconnection would provide a starting point for an industry "which has spent virtually all of its life to this point, in a protected, slow moving environment, without somewhat less than the desired degree of innovation."

Other departments such as Treasury Board were insistent that permission for interconnection with the public-switched network was essential. "It is a federal government objective," one Treasury Board document stated, "to see established a national system of public data communications facilities for use in the provision of or access to, computer or data communications services by any segment of society subject to appropriate rules and conditions governing entry and participation." By 1973, Gérard Pélletier could announce that the dialogue over interconnection was "thick and heavy"; but any proposed changes in the rules of the game had to be considered very carefully both at the national and at the regional and provincial levels.[41]

The issue of interconnection was a divisive matter for the provinces, and was symbolic of the problems that provincial communications ministers had in reconciling vastly different experiences and interests. Ontario's position favoured a more competitive environment, with Quebec a little more reluctant to break the principle of local monopoly, and Alberta and Saskatchewan dead set against any liberalization of interconnection policies. Ontario was supported by a number of small independent suppliers eager to crack the multi-million-dollar market for interconnection equipment and private communications equipment.[42]

Officials within the Ontario MTC viewed Bell's handling of the interconnection issue with some cynicism. Robert Bulger, the ministry's most trenchant critic of Bell's practices, viewed the company's attitude as exploitive. After the American Federal Court had decided on the Carterfone case, a number of private suppliers in the United States flooded the market with terminal equipment offerings, often without regard to circuit compatibility or a "firm knowledge of coupling devices" required to assure the smooth operation of the network. The US Bell system reacted by demanding conditions that would, in Bulger's view, delay interconnection and protect their terminal equipment market. "In summary," Bulger argued, "*phased* interconnection is viewed to be in the public interest. We must now address ourselves to the 'rules of the game.'" Ontario estimated that over 100,000 illegally connected pieces of equipment were being used in Canada, the preponderance of which were telephone answering devices. Although MTC officials maintained that Bell should continue to provide offerings throughout its territory, interconnection was a "fact of life and desirable" from the consumer standpoint.

By 1974, the Ontario minister of transportation and communications could definitely state that he was "cautiously in favour" on the subject of possible liberalization of the tariffs relating to the interconnection of customer provided terminal equipment in Canada. "However," John Rhodes warned in a letter to an interconnection supporter, "I would reiterate my concern about the impact of interconnection upon the Canadian telecommunications manufacturing industry and hope that any steps taken by the federal government will be to move carefully in this area." What was therefore required from Ontario's point of view was the protection of jobs in the telecommunications manufacturing industry in Ontario, as well as some form of protection against increased costs to the basic telephone subscriber in Ontario. The province favoured a "phased de-regulation" in order to mitigate any negative impact on the user, and especially residential subscribers who were "generally satisfied with the status quo."[43]

Perhaps those most opposed to the liberalization of interconnection rules were the Western provincial communications ministers. "Through its shortsightedness in pursuing its own objectives," one stiffly worded statement ran, "the federal government has failed to perceive the essential need for a monopoly environment in the provision of telecommunications services." On the issue of interconnection in particular, the western provinces were particularly concerned that a permissive interconnection policy would "have damaging effects upon the provincial telecommunications networks and services provided to their customers." The prairie provinces feared that letting privately-owned terminal devices attach to the public network would increase costs at the expense of the general telephone user. In other words, provincial utility interests determined provincial interests.[44]

By 1975, provincial officials were complaining that interconnection was moving too fast, had little or no federal legislative framework, and seemed to be diverted by Bell Canada. At one meeting between Ontario, federal and Bell officials, the Bell representatives argued that the provincial input was unnecessary. "Why worry about the provinces, why don't you people [the DOC] implement it immediately?" Bell's statement was likely embarrassing to the DOC, but the provocative tone signalled to Ontario officials that Bell was ready to move on interconnection, "and within their time frame."[45]

Perhaps the clearest statement of the provincial position on interconnection came in mid-1975 in *Canadian Interconnection Magazine* where

Ontario's director of telecoms, Robert Bulger, announced that as far as Ontario was concerned "no tight definitive policy exists at this time, and purposely so. We felt that Ontario would have to feel its way and be flexible enough to adjust to changing needs and demands on both sides in order to rationalize a policy on interconnection."[46]

The Quebec position on interconnection, seen from the perspective of the late 1970s (when documentation is available), was that the question was essentially political "plutot [qu'] une question de philosophies de developpement des communications."[47] The provincial government opposed the interconnection of CNCP and Bell because of the effective creation of a duopoly. The reason was not hard to understand: in order to assure the harmonious development of telecommunications in Quebec, "le jour où les activités de Bell Canada au Québec serons sous control reglementaire québécois" should be anticipated and protected.

Pressures for interconnection forced telecommunications interests and federal and provincial governments to stake their claims in a concrete fashion. That process revealed the fundamental divisions not only between levels of government, but also between players in the industry. In the meantime, Bell Canada bided its time and prepared strategies to retain its dominant control over the network. The resulting disagreement symbolized the difficulties of reaching consensus amid a multiplicity of competing interests.

CONFLICT AND DISAGREEMENT

The process of federal-provincial discussions over communications policy continued in 1974 and 1975 but failed to reach any agreement. They had ended in deadlock between a national government determined to protect its jurisdiction and control over telecommunications, and exasperated provincial governments convinced of the necessity of a greater provincial role.

As the last section showed, Ontario officials favoured interconnection. They were willing, however, to lobby relatively quietly and not press the federal government unduly over the issue. The province's strategy on interconnection was in many ways symbolic of its position over broader communications issues in general; it reflected the province's virtual ambivalence on some issues, and vague positioning on others. As Ron Atkey, political advisor to the premier noted, "there has been no concrete evidence that Ontario is prepared to move substantively in the communications field." Atkey cautioned that Ontario's lack of action likely produced a feeling in Ottawa that Queen's Park need not be taken seriously. Atkey advised the MTC to introduce legislation, if only to clarify its goals and objectives in the field. Only then might the federal government pay attention.

Atkey's analysis of Ontario's position in communications was not far off the mark. The provincial government simply had not made a convincing case to the public why provincial involvement in telecommunications was so important in the first place. Its position on common carrier regulation was confusing, and its reluctance to forge a united front with other provinces was damaging to interprovincial efforts to bring Ottawa to the bargaining table. Atkey suggested that the province was favourably positioned in telecommunications, especially since the CTC had proved itself virtually incapable of regulating Bell and satisfying the consumer interest. Sooner or later, Queen's Park would have to convince the federal government of its serious intentions in the field. Atkey was blunt: the Ontario government should get a foothold in the field, or quit and concentrate on other areas of government.

Atkey also suggested that Bell Canada should be persuaded that federal regulation did not serve its corporate interests as well as provincial supervision might. Atkey offered the argument that the grass just might be greener on the provincial side of the fence. If the company supported provincial ambitions in the field then Ontario might undoubtedly be able to do something for Bell.

Ontario's position in the growing debate had been compromised by the strength of animosity between Ottawa and Quebec City, and especially between the two ministers, Gérard Pélletier and Jean-Paul L'Allier. When a tripartite meeting between Rhodes of Ontario, L'Allier, and Pélletier was suggested, one Ontario advisor reminded Rhodes of the history of personal animosity between the two Communications ministers "of a kind that is hard for English Canadians to understand." What was more, Ontario would be put in the role of intermediary between Ottawa and Quebec City, since their respective positions were furthest apart, leaving Ontario little room to advocate its own position. If news of the secret meeting ever got out, the damage that news of a central Canadian solution would bring to interprovincial solidarity could be incalculable.[48] The difficulties over communications issues were made even more complicated when considered in the larger struggle to redefine federalism and the constitution.

For the short term, Ontario would opt for a big push on the question of jurisdictional responsibility, which was met by continued resistance on the part of the federal government. Rhodes's hard line approach – to open the jurisdictional question "come hell or high water" – was part of a strategy to get federal-provincial talks moving, and was not necessarily critical to Ontario demands. The Ontario minister informed his federal counterpart that there was no useful reason to hold any talks as long as Ottawa was not ready to talk about jurisdiction. Rhodes indicated the willingness of the provinces to move ahead with a federal-provincial conference, even, oddly enough, without the federal government in attendance. Rhodes even suggested to Pélletier that such a meeting could be held in Ottawa, "with the provinces directing their remarks to an empty chair."[49]

Rhodes' suggestion reflected his frustration. The Ontario minister had obviously taken Atkey's advice and raised Ontario's profile in the debate over communications. At the same time, however, he was expressing the provincial frustration at being utterly unable to move the federal government from its national telecommunications plans, even at the risk of a complete breakdown in federal-provincial discussions.

The federal communications minister publicly expressed his frustration with provincial delays after the November conference of the pervious year. "Despite our preference for a negotiated solution," Pélletier wrote to the provincial chairman, "if we cannot set aside the jurisdictional dispute, recourse to the courts may become inevitable." Pélletier told the parliamentary Standing Committee on Transport and Communications some time later that his department had wished for a conference in February 1974, but the provinces, who had to work out their respective positions and present a united front, refused.[50] The road to the second federal-provincial conference, scheduled for May 1975, was thus a difficult one to travel. Negotiations continued throughout the winter of 1975, but little signs of consensus emerged. One month before the conference, the federal Department of Communications released its grey paper on communications policy entitled *Communications: Some Federal Proposals*.

The paper was at least partially the product of federal exasperation at their inability to obtain agreement on a national communications policy. In order to give "comprehensive attention" to the development of the Canadian telecommunications system as a whole, the federal government "expected that provincial governments will be willing to enter into the proposed arrangements in a cooperative spirit." The federal government stood firm on the issue of jurisdiction: any formal transfer of legislative authority was ruled out.[51]

As an alternative, federal-provincial cooperation would be encouraged through a variety of consultative mechanisms. The arrangements included the establishment of a Committee for Communications Policy at the ministerial level. On the regulatory side, the federal government reaffirmed the transfer of telecommunications responsibilities from the CTC to the CRTC. Second-stage legislation at the federal level would then establish "a coherent body of federal law on communications" with the governor-in-council retaining the preponderant share of decision-making power. Pélletier also proposed an Association of Communications Regulatory Bodies to institutionalize provincial input. In order to meet the concerns of Ontario and especially Quebec, representatives of the regulatory bodies of those provinces would be entitled to participate in the proceedings of the federal commission before it rendered a decision. As a trade-off, the federal government expected the right to participate in hearings by the regulatory bodies of Ontario and Quebec on matters that had extra-provincial aspects. The message from Ottawa was clear: the federal government was determined to move after five years of study to recast telecommunications regulation, whether or not the provinces were willing to cooperate.

The federal government thus came to the conference with an attitude hardened by years of disagreement and lack of progress. The provinces appeared to be united after some coaxing at a meeting of provincial communications ministers by Jean-Paul L'Allier at Lac Beauchamp. L'Allier explained that the meeting "a pour but de faire le point sur les divers items a l'ordre du jour afin de constituer si possible, un front commun de toutes les provinces a l'égard des propositions fédérales." Pélletier countered by hoping that "Canadian" objectives would be met. The least Pélletier hoped for was a Council of Communications Ministers to discuss communications issues in the future. Discussion of the redrafting of the British North America Act was out of the question. This time, however, any potential failure of the conference would not deter the federal government in its aims to revamp telecommunications regulation. In Quebec, the press spoke of the "conférence de la dernière chance." It noted that L'Allier "bataille depuis 1971 pour tenter en vain maintenant de repatrier au Québec les pouvoirs de décision ... le gouvernement Bourassa a un besoin absolu, s'il veut effectivement proclamer la souveraineté culturelle d'ici le prochain test electoral." The failure of Quebec at the conference would mean a body blow to the concept of cultural sovereignty and to Premier Bourassa's attempts to reconcile the federalist and sovereignist impulse in one policy.[52]

Pélletier did manage to break the stalemate at a dinner meeting of ministers and deputies and dissolve the common provincial front. Quebec found itself isolated while all the other provinces save British Columbia agreed to accept sole federal jurisdiction over telecommunications and broadcasting. The provincial consensus built by L'Allier's efforts had collapsed. "Notre place dans la confédération ratatine de jour en jour," L'Allier bitterly remarked. The communications dossier would have to be examined at the level of the first ministers. At that level, Prime Minister Trudeau's objectives on the repatriation of the constitution conflicted with Quebec's cultural sovereignty initiatives, in which Quebec control over communications was crucial. Deadlock seemed inevitable. René Lévésque and the Parti Québécois were quick to sound the death knell not only for "le mythe de la souveraineté culturelle" but also for the "front commun des provinces canadiennes."[53]

The federal government's refusal to negotiate over the contentious issue of jurisdiction, as much as provincial insistence that jurisdiction was a critical issue, doomed the July conference to failure.[54] Provincial solidarity had crumbled in the face of adroit negotiation on the part of the federal government, and the offer of the establishment of a new Council of Communications Ministers to achieve consensus. The most the provinces could hope for was federal delegation of authority.

The technocratic dream of a comprehensive, orderly national telecommunications policy resulted in complete failure by 1975. Ironically, the very comprehensiveness of the proposals, and the fact that they were too close to the status quo heavily favouring federal jurisdiction, forestalled any real possibility of federal-provincial agreement. Equally, the conviction at both levels of government that jurisdictional control was the only means of ensuring state interests also doomed national initiatives.

Quebec had nothing to lose from a confrontation with the federal government over telecommunications. Indeed, the disagreement strengthened nationalist elements within the governing party, not to mention the province. By pushing a maximalist policy and refusing to negotiate on specific issues, Quebec effectively precluded more pragmatic sanction that would have seen greater Quebec input while acknowledging legitimate federal aspirations in the maintenance of a national telecommunications system.

For its part, Ontario was caught in the middle. Making common cause with Quebec was natural: both provinces had negligible control over telecommunications in their respective territories. Strategically, an Ontario-Quebec alliance over communications issues also made perfect sense. A common provincial rights front was a tried and true technique for exerting pressure on the federal government. However ambivalent or annoyed Ontario professed to be over federal communications proposals, provincial officials were not sufficiently motivated to seek total control over telecommunications. Not animated by an ideological attachment to control, Ontario would be satisfied by a pragmatic compromise in the field. This included the reform of federal institutions, as long as there existed a sufficient quid pro quo. For the province, the defence of the provincial interest could be accomplished within the prescribed parameters of federalism.

For the federal government, the failure of its attempts to erect a national telecommunications policy demonstrated the limits of the marriage of federalism and technocracy. The fusion of a decentralized polity on the one hand and attempts at central technocratic control on the other resulted, not surprisingly, in stasis. Intergovernmental conflict had diverted crucial attention from the pressing problems of regulatory reform in the government's own back yard. Discussions continued after 1975, but not again with the same intensity. The strife had left the federal state drained of energy, and telecommunications policy reform assumed a lower priority. Instead, the policy vacuum was filled by ad hoc decision-making and by the new powerful federal regulatory authority, the CRTC.

The real winners in the failure of telecommunications discussions between 1970 and 1975 were companies who would stand to gain from the preservation of the status quo. This meant that companies such as Bell Canada would have more time to position themselves in the marketplace for eventual changes. In the meantime, however, companies which did not benefit, such as CN/CP Telecommunications, would have to await regulatory remedy in the late 1970s before relief could be given.

7 "Abort, Retry, Ignore, Fail?" Ottawa and the First Information Highway, 1969–1975

Our world is mostly directed by groggy bureaucrats who decide to make innovations, such as satellite environments, without an inkling as to the social and psychic consequences. It is not possible to point directly to a single group of twentieth century individuals, however small, that is governing a business, or a nation. The show is run by nineteenth-century people.

– Marshall McLuhan

I felt that no attention was being paid to the long run challenges facing the country, and that many of our ad-hoc decisions could only worsen problems ... I have always been convinced that we are our own worst enemy ... The officials here [in Ottawa] are completely out of tune with the country and the ministers ... have no time for thinking and follow the easier course of going along.

– Eric Kierans to Walter Gordon, 1971

Nobody could have accused Marshall McLuhan and Eric Kierans of coming from the same intellectual or political backgrounds, or of sharing common views on most matters. Kierans was the business-man-cum-politician, McLuhan the aphoristic prophet of the electronic age. These two strong communications celebrities came from different worlds, and so it was all the more remarkable that they would come to similar conclusions about state intervention into technological matters. The difference between McLuhan and Kierans came in their appraisal of the ability of the state to effect change. Kierans believed that a more rational administration sensitive to public feeling could harness the national government for maximum local benefit. McLuhan believed that the hopelessly anachronistic mindset exemplified by Kierans and his colleagues would lead more to confusion than create innovative responses to technological change.

By the late 1960s, the most portentous technological change in the minds of policy-makers was the incipient fusion of computers and communications. Although the possibilities of time-sharing systems were recognized by the late 1950s, it was only a decade later that the technology became significant, allowing terminals to be placed at user installations to provide on-line access to a mainframe computer. Raw computer power could be used for commercial applications, and at the manager's point of contact as well.

Important innovations generate questions. New technologies routinely cause states and societies to question how the innovations should be shaped, and by whom. In Canada, computer communications held such a potential. Policy-makers believed that answering those questions would require a political transformation in the power of government. Remarkably, this fusion of these two technologies sired a powerful force that led the state to attempt to substitute public policy for the decisions of the market. In this sector at least, the market was deemed to be deficient in reflecting the values of national sovereignty, indigenous technological development, and maximizing economic growth. The notion that computer communications had common property qualities produced further doubt that the market could diffuse its benefits. Computer communications would produce a future Ontario Hydro, not a BCE.

This chapter will examine the extraordinary efforts of the Canadian state to deal with the advent of a promising new technological development between late 1960s and the mid-1970s. Policy-makers in the later 1960s were true believers in technology's power to reorient political and economic relationships. They were also convinced that the state should participate in the distribution of technological benefits. These convictions led them to strange waters indeed. However, the divergent interests of the players – common carriers, domestic and US computer industries, and provinces – would put that conviction to the test. What became obvious by 1975 were the strict limitations on state action imposed by circumstance, will, and inertia. But in the late 1960s, everything seemed possible. And so it was that federal policy-makers embarked on building the first information highway, twenty-five years before it became a reality.

AN EMERGING CONVERGENCE

By the 1960s, technological innovations in computers and communications sparked discussion of an equipment utility that would fuse the two technologies into a national network. Owning a large computer would not be necessary. Western Union advertised their system in the United States as "designed to provide information, communications and processing services in much the same way as other utilities supply gas and electricity."[1] In Canada, this development struck a resonant chord, for two reasons. First, the Canadian experience with utilities had imprinted the state's thinking about the significance and potential uses of the new technologies. On this terrain, the familiar could meet the future: the public utility could meet the computer. This alliance of electronics and communications created an opportunity for patriating a technology that had been almost wholly imported. "Five years from now," the *Financial Post* reported in 1967, "many Canadian companies may buy their computing power just as they buy their electricity or gas today – from a utility supplying service from a big central computer … The day of the computer utility is here." Perhaps the day of the computer utility would belong to Canadians as computer technology had not.

The imminent arrival of the computer utility was given further currency by science. J.W. Graham of the University of Waterloo proclaimed in 1967 that the technology had reached the point "where it is realistic almost immediately to have public-utility type of computer. In fact it is a reality." Graham believed that one big computer with universal access was not only possible, but also desirable.[2]

In Toronto, both Burroughs and Canadian General Electric were offering time-sharing services on large systems, and experts were predicting that inside five years over half of all computer work would be done on a time-shared basis. The reasons were simple: instant availability, user flexibility, and access to an untapped electronic power supply via the existing telecommunications infrastructure. IBM provided a service based in Toronto called Quicktran, mainly used for scientific computation. The "forerunners of tomorrow," computer utilities were at last becoming economically significant.[3]

Speaking to Canadian computer scientists in 1966, Robert Fano of MIT warned that these new utilities would trigger the interest of the state, especially over questions of ownership and control. He warned that the worst thing that could happen to technological development in the field would be resolution of questions of control without a proper and well-informed debate.

Fano's predictions were valid. It was in the spaces between telecommunications and computers that Canadian policy-makers would seek an opportunity to carve a niche for a national policy. The results of failure were said to be great: technological impotence, the result of a complete dependence on foreign suppliers for computers. The "Just Society" might not sustain itself if the Canadian market could not compete in this high-growth area.

The event that provided the entry point for federal government intervention was the takeover of Computer Sciences Canada by CN/CP in March of 1969. Both Canadian National and Canadian Pacific each paid half a million dollars for controlling interest of the Canadian subsidiary of Computer Sciences Corporation of Los Angeles. The acquisition was made to promote data communications and establish a stronger national linkage between capital and technology in a Canadian-based computer service company.[4]

The details of the acquisition were innocuous enough, but the implications were viewed as serious and instantly recognized as such by the telecommunications and computer titans. IBM did not view the development as a threat, but took pains to stress that any cross-subsidization should be prevented. Bell Canada protested that it was shut out of the new business owing to the restrictions imposed by Parliament on the company's new charter which had prevented the company from entering the computer service business.

The intention to deal with communications issues had assumed a new vigour the year before with the creation of the Department of Communications. The government was quite aware that the information retrieval would be an important component of national technological progress. Internationally, this information explosion was based on the availability of increasing telecommunications services.

A commitment to fostering technological capacity existed, but it had to be translated into difficult decisions about industry structure and state involvement. The opportunity for the state's new-found resolution about technological progress would come in relatively rapid succession between 1969 and 1974. Those five years were the window of opportunity for the federal government to put its imprint on technological development. Once distinctly separate areas, telecommunications and computers were becoming intertwined, forcing the government to consider the implications. What social and political product would emerge from the regulation of these new econotechnical systems? What role would the Canadian government have?

Communications officials quickly came to the realization that regulatory powers were utterly inadequate to manage new telecommunications services. In telecommunications, the Railway Act did not regulate private wire services. Yet the demand surge in data transmission and other telecommunications services were occurring precisely along these wires. "The whole relationship," Eric Kierans told the Standing Committee on Transport and Communications, "between telecommunications and computers is in a fluid and changing state and is a matter of great importance for the economic growth of the country, its social structure and our relations with our neighbours."

The state had therefore not staked its regulatory claim in an area of considerable importance and activity.[5]

The loophole was a critical one. The department anticipated that, within a decade, computer service centres would provide computer power "in a manner analogous to the distribution of electric power." The systems themselves would establish an information processing and distribution grid that would handle anything from reservations to medical diagnoses.

The department was convinced that the expansion in the computer utilities field would continue for some time to come. It would bring data processing and telecommunications into "unprecedented intimacy," and induce Canadian common carriers to attempt market entry into a new utility market. The acquisition of Computer Sciences (Canada) Limited (CSC) by CN/CP, approved by order-in-council in January, signalled that the transformation was at hand. The CN/CP purchase caused consternation among the computer utilities, who claimed that CN/CP could use their market advantage to squeeze the threatened firms out of the market.

The issue was a complex one, but the Department of Communications felt that at the very least private wire services must be regulated. The department assured that "CN/CP will not be in a position to cross-subsidize their computer operation by tendering advantageous contracts for telecommunications lines with this subsidiary." In the meantime, Ottawa was warning provincial utilities such as Québec Téléphone to refrain from entering the computer time-sharing field until the issue was resolved.[6]

The intention to regulate private wire services was an interim response to the problem raised by the computer services industry. Allan Gotlieb was determined to prevent CN/CP from charging one price to its subsidiary and another to the market.[7] A comprehensive state response to the larger question of carrier participation in data processing, however, would have to wait further study.

The elements surrounding the purchase of CSC neatly typified the Canadian technological dilemma. CSC's American parent firm, Computer Sciences Corporation, was one of the giants in the information science field and developed the software for its Canadian subsidiary. The only way that CN/CP could readily enter the market was to acquire the subsidiary and contract for the technology. One option under consideration to forestall the continentalization of the computer/communications field was the use of the backers of the Trans-Canada Telephone System, principally Bell Canada, to provide competition in the computer utility field. But that cure might be worse than the disease.

This was the dilemma presented to cabinet in the spring of 1969. The marriage of computers and telecommunications would create a massive advantage for telecommunication companies who controlled the means of transmission. The advantages were tempting: large pools of raw computer power would lead to reduced cost and prevent American penetration by consolidating east-west links. One cabinet document warned that "computer utilities are developing much in the fashion that power utilities evolved fifty years ago." In addition, the new utilities promised to have a phenomenal rate of growth and become "as much a part of everyday life as the telephone is today."[8] Would these utilities be owned and regulated in a similar pattern to the telephone system or electricity grid? It was too early to tell. Further, cross-ownership could spur innovation in switching, and terminal apparatus, not to mention the diffusion of these technological innovations. The consequences of fusing monopoly telecom control and computers might have disastrous consequences.

The purchase of Computer Sciences Canada Limited elicited a flood of advice about how to handle the apparently rapid formation of information utilities. In July 1969 several computer service companies sought to focus the federal government's attention on the "serious implications" that had issued from the CSC-CN/CP deal. The acquisition had been given cabinet approval shortly after the filing with the Ontario Securities Commission. Companies in the field such as System Dimensions Limited, Computel, and I.P. Sharp Associates were concerned that the approval had pre-empted public policy formation, especially when the issues were not clearly understood. Construction of an electronic highway had serious implications for national economic and technological progress. If the possibilities were staggering, so too were the pitfalls. Canadian imports in the electronic field demonstrated the borrowed nature of the country's technological advance. Canada's "technological balance of payments" exposed the peripheral nature of the country's technological modernization. The US–Canada balance was 7:1 in 1956 and 9:1 in 1965.[9]

The presence of IBM cemented the domination of foreign technology. Its position over the electronics capital goods market led the Canadian computer service industry to suggest that counter-attack would be unavailing. It would only serve to increase cost to the domestic user. Paradoxically, the industry was also convinced that the technological mantle fell upon them, at least in certain areas. A dynamic domestic computer technology industry should be allowed to develop the field with state support. The industry argued that technological specialization, combined with stimulation of native industry, could turn the situation around. The marriage between communications common carriers and the computer service industry should therefore be annulled. The only hope for the Canadian situation was the removal of the common carriers, who were often laggards in technological advances. The greatest challenge to public policy-makers was the maintenance of competition and innovation. Barring common carriers from the computer services industry, they argued, would inoculate a strategic industry against the unwholesome effects of monopoly control or corporate concentration.

Meanwhile, Kierans and his department encountered some difficulty in combating the notion that the federal government was about to raise billions of dollars to nationalize the private telecommunications companies, and then hand over the computer industry to the common carriers. After assurances this was not being planned, Kierans nonetheless reaffirmed the desirability of a national computer network: "computing power distributed to customers from large central installations through communications systems [is] an important development that could have great economic, social and political impact on Canadians collectively and individually."[10]

The DOC turned to the computer industry leader, IBM, for advice. Change and convergence in computers and communications required "timely and informed resolution" of the problem in order to ensure the orderly development of the respective industries.[11] IBM responded that regulation could only hinder the sales capacity of computer services and concluded that "there is no need for regulation of data processing services, and regulation would have undesirable effects."[12] Nevertheless Kierans was concerned above all with an effective state response to an economic and technological development of the first order. For a parallel, policy-makers turned to the historic development of the Canadian utilities sector. "While I recognize the validity of some of the objections to the term 'computer utility,'" he suggested in 1970,

it still seems to me to be one of the best description of what appears to be happening on this continent. And what is happening ... is something analogous to the development of the huge hydro and communications utilities which took shape at the turn of the century and now link all our homes, factories and offices even in remote villages and hamlets.[13]

The crucial difference between the old and the new, he argued, was the incredible speed with which the computer utility was forming, perhaps because it built upon an established continental communications grid. Whatever its complexities, national control had to be protected. "If we in Canada lose control of this essential information industry," Kierans warned, "any effort to maintain a distinct Canadian political entity would be futile." The international dimensions would also be of decisive importance. The need to develop a political philosophy to deal with developing communications systems was of the utmost importance.

Kierans could take comfort in the winter of 1970 that a genuine computer utility industry was indeed developing, linked with the communications resources of the common carriers. But at every turn and in every public appearance he took pains to stress the dangers of continental integration in computer communications. Canadian anti-dumping laws did not apply to US exports of data processing capacity, intensifying the danger.[14]

The same day Kierans was publicly warning of the dangers of American penetration, the cabinet finally met to consider participation by telecommunications carriers in public data-processing. If participation were granted, under what conditions would the carriers operate? But the cabinet was being asked to decide on a much broader question. Resolving the problem of carrier participation would also mean determining the pattern of a new utility which "promise[d] to make the computer as much a part of everyday life as the telephone is today."[15] The decision could not wait for the Telecommission's report on computer utilities expected in 1971; matters were moving too quickly.

The cabinet met after considerable pressure was brought to bear on Kierans and the minister of industry, trade and commerce, Jean-Luc Pépin. The telecommunications common carriers, especially Bell, were anxious to enter the field. CN/CP already had, in January of the previous year. Since then, the company had been developing an aggressive promotion of the utility. The independent data-processing industry feared the worst.

The situation as it stood was not reassuring for those anxious to promote domestic control of the industry. The two giants in the computer utility sector – IBM and General Electric – were both controlled from the United States. From that country flowed the technological innovation, the expertise, and the capital to sustain and diffuse the new utility. Kierans and Pépin were troubled by a situation that "could result in almost total dependence on computer utilities owned and controlled in the United States."[16] On the other hand, Canada's telecommunications utility was largely locally controlled and had coast-to-coast systems that could be basis for an all-Canadian utility distribution route.

Kierans and Pépin placed the causes of the continental drift directly at the door of government policy. First, the electro-magnetic transfer of information was tariff-free. Second, anti-dumping restrictions did not apply to the leasing of surplus data processing capacity at less than cost. Third, Canadians were paying at least 20 per cent more for computers than Americans because of exchange, tariffs, and sales taxes. Of the $160 million in imported electronic computers in 1969, only one-quarter passed free of duty, and customs revenue from computers alone exceeded $14 million. If the trend continued, aided by what Kierans saw as a disastrous excise policy, the computer utility would go the way of the auto industry, with information processing resources anchored to a few immense organizations in the United States. One need only examine the government's own procurement to see where the trend was going, as table 23 indicates.[17]

Kierans and Pépin reminded their cabinet colleagues of the grave implications for Canadian sovereignty, and that the country was historically unwilling to submit to the impact of market forces on essential links. "It is not necessary," a cabinet memorandum explained

to underline the grave implications in terms of Canadian sovereignty. Historically, Canada has been unwilling to submit to the unconstrained effects of market forces on its essential services. In the development of railroads, telecommunications, highways and air services, the vital importance of establishing an east-west axis has been recognised at some cost. A parallel can be drawn with the present need for a policy that will facilitate the establishment of a national Computer Utility network owned and controlled in Canada.

Innovation had sired a new utility. Policy would have to coax it in a Canadian direction.

The belief that this computer utility was at hand was reinforced by the idea that economies of scale existed. Large computers were faster and more flexible than smaller machines, not to mention less expensive per task. The new utility would capitalize on these advantages, and boost machine efficiency and use.

The state had to hardwire the country before transnational lines were imposed on it. Time was not on its side: the cabinet looked at the developments in wide-band telecommunications, satellite technology, and digital modulation techniques and saw global information utilities that would "negate plans and policies based on narrow nationalistic considerations." The trend in the meantime was towards concentration and largesse. The cabinet thus had to link questions of Canadian ownership and control with problems of equalization of access, common use of equipment, and optimum system design. CN / CP's claim in the data processing field with the purchase of CSCL would also prove difficult to undo.[18]

For help with the perplexing fusion of telecommunications and computers, policy-makers recalled the changes made in the structure of railway transport and trucking in the 1950s. The analogy of railway diversification into the trucking business was felt to be especially relevant. The royal commission on the subject concluded that the fears that railway entry would end competition in trucking transportation were baseless. Careful regulatory oversight would ensure that neither cross-subsidy nor discriminatory treatment would occur. A similar formula could be transposed to computer communications.

To be sure, interweaving computers and telecommunications would mean a lot more work for the regulatory agency. A horizontally integrated computer utility that offered raw computer power for communications as well as applications would be a regulatory gordian knot. The CTC had enough trouble with a straightforward breakdown of telephone costs. To expect them to master the intricacies of the marriage of telephones and computers was beyond hope. The "level of sophistication and control in communications regulatory bodies," one official noted, "is not currently available in Canada and could not be created overnight."

The cabinet was also aware, but not entirely convinced, of the computer utility industry's arguments that carrier entry would ensure a slow pace of innovation in new techniques and devices. For its part, Bell noted that data communications comprised but 5 per cent of total demand, and innovation was introduced at a pace that took into consideration the interests of all class of customers. The danger existed that public data processing would weaken their commitment to their fundamental obligations to universal service.

Added to the questions of industry structure was the wild card: federal-provincial relations. By the early 1970, provinces were awakening to the potential of data processing services in harness with telecommunications. If provincially controlled carriers were allowed to participate in data processing, cabinet's careful calculations would become a waste of time. It might be forced to act quickly to protect the carriers operating under its control.

The cabinet was thus faced with the challenge of giving a new utility its pattern, if not its fundamental thrust. This new utility could be the anchor for a Canadian computer industry, strengthening the export potential of the telecommunications sector as it developed. The political dimensions were not as promising. Allowing carrier entry would be interpreted as capitulating to pressure from Bell, and cutting loose the data processing companies that worked on the margins. To exclude the carriers and the government would be selling out to IBM and General Electric.

Yet exclusion of the carriers from data processing would reinforce the continental integration of computer communications services and eliminate a powerful national champion. The independent Canadian data industry would be vulnerable to the depredations of the titans such as IBM and General Electric as well as the American independents who viewed Canada as a natural and logical extension of their home market. As matters stood, Canadian computer companies controlled no more than 10% of the Canadian market.

Support for carrier entry came from the manufacturers of computer equipment and large-scale users, including the pulp and paper and oil-refining industries. Computer utility companies, computer leasing companies, and the chartered banks were opposed. But many also expressed the view that a reciprocal relationship should exist. As far as Consolidated-Bathurst was concerned, "there seems little point in permitting a small restricted number of common carriers to offer Data Processing Services without permitting Data Processing Organizations to offer communications services."[19] Indeed, when asked to reply to a government questionnaire on the subject, many of the key corporations favoured the entry of data processing organizations into communications services. IBM also favoured such a structure, but reminded the federal government that a data processing company which used a common carrier to provide communications ancillary to data processing should not be regulated.

The response of the common carriers to even the hint of reciprocal provision of data processing and communications was predictable and fierce. The TCTS insisted that proliferation of telecommunications facilities would result in duplication, interrupt emergency communications networks, and undermine economies of scale. The organization argued that the public would be ill-served by having to deal with a number of companies for its communications needs. "The future need is for simultaneous voice, data computer and visual interaction." CN Telecommunications also argued that the present number of carriers was more than enough for the demands of Canadian business. CN/CP's subsidiary, Computer Sciences Limited, responded this way: "The total memory capacity of the approximately 2000 digital computers installed in Canada is estimated at 500 Megabits. One dedicated microwave link with a total transmission capability of 25 megabits/second would be sufficient to exchange all information stored in the memories of all computers in Canada once every 20 seconds. This capability is orders of magnitude greater than any predictable requirement in the data processing industry."[20]

The division of opinion along the lines of the interests of users was encapsulated in the response of the Iron Ore Company, which provided two contradictory responses to the question. As a common carrier, the company favoured the joint operation of data processing and communications. As a data processor, it argued that non-carrier data processing enterprises should be barred from providing data processing services, warning of the consequences of overcrowding spectrums and duplicating facilities.

The problems of data processing and telecommunications, and the enmity between the two industries, was fostered by the asymmetrical relationship between them. The common carriers argued that there were no significant irritants in the relationship, and boasted that the technology of the Canadian carriers was second to none. They reluctantly admitted that better cooperation was needed between users, computer suppliers, and the carriers, but were generally satisfied at the progress of the technology.[21]

The much smaller computer services industry was almost entirely dependent on the telecommunications companies, mainly the common carriers. The report recounted the feeling of oppression the industry felt at the hands of the common carriers. Bell's attitude was often hostile to the computer services companies. A punitive rate structure was squeezing margins, and transmission facilities were lagging behind the state-of-the-art.[22] The restrictions on use of modems and the problems of interconnection of common carrier leased and switch services with privately owned communications services was another irritant.

However, both sides were careful to stress that the "utility" in "computer utility" should not carry with it the usual regulatory implications. Indeed, there were sound reasons to consider this utility un-utility like: capital requirements were low, the ratio of revenues to capital investment was high, and the types of services it could provide were varied. Additionally, while the technology permitted certain economies of scale, software limitations placed a ceiling on size of operation.

In any case, industry on both sides of the issue paid homage to the protection of the "competitive market" for data processing. This was certainly ironic given the dominance of IBM on the data processing side and Bell on the telecommunications side. But the pre-empting of intrusive government regulation was a concern shared by large and small. State regulation would retard technological progress in the field and mar industry flexibility. Most were agreed, however, that the state should regulate security and privacy concerns, as well as standards.

IBM persistently expressed its concern about the government's use of the word utility. It sought and got assurances that the statements made by the government were intended to "explore ideas and to generate debate." The corporation was careful to stress their dissatisfaction with the entire concept of the computer utility, which it perceived to include total regulation of the field. Allan Gotlieb took pains to assure the company that there was no "fixed or 'master' plan already in existence" regarding the concept. Still, it was evident that his department believed there was some urgency about controlling the computer industry for Canadian ends. "It is up to Mr. Kierans," the *Toronto Star* warned, "to see that computer information in Canada, like the railroad lines, flows mainly east and west, not north and south."[23]

The issue was then passed to the Economic Policy Committee of the Cabinet a month later. The federal government would not exclude at least the consideration of allowing telecommunications carriers to engage in public data processing, primarily for reasons of national economic development. At the Privy Council Office, R.B. Bryce concluded that if the current data flow situation continued unchecked, it could pose serious problems of sovereignty over information stored in US facilities.[24] Ultimately, however, the federal cabinet decided that too much concentration would result in allowing telecommunications carriers to engage in data processing. As a result, the carriers were ultimately denied entry into the computer communications and data processing fields.

The decision did not come without a fight. Simon Reisman, the Treasury Board deputy minister, was convinced that denial of common carrier entry was the wrong decision. He stated the case for convergence, arguing that "in the communications and computer industries, the common carriers are probably the only Canadian firms with sufficient resources to resist the pressures to develop north-south networks based on the use of computers installed in the United States."[25] Reisman warned that the pressures were substantial and related to the absence of tariffs on the electro-magnetic transfer of information. A situation such as that would "pose major problems for sovereignty," for it was theoretically possible for the United States Supreme Court to order information stored by Canadian firms be made available to American governments.

Reisman also believed that the pressures created by CN/CP's acquisition of Computer Sciences Limited in 1969 had led to a hastily conceived policy. He argued that application services should not be provided by the carriers. Reisman believed his course of action would cause the least harm to the established Canadian computer utility companies, and make the combined field easier to regulate. Only by horizontal diversification should the carriers be allowed to offer the services.[26]

By letting the carriers provide raw computer power through horizontal diversification and application services through autonomous subsidiaries, Reisman concluded that Canada would avoid the perils of American technological encroachments. Those perils would have to be fought some other way.

CRISIS AND OPPORTUNITY?

The question of carrier entry was only the first of a series of technological challenges encountered by the federal government. As in telecommunications regulation, the state became the arena where provincial governments, utilities, and private industry struggled for advantage. Unlike the more established telecommunications regulation, however, computer communications would require the development of a policy with few of the traditional instruments or little of the political salience of regulating telephone rates.

With the question of carrier entry resolved for the moment, the Department of Communications moved to flesh out a comprehensive policy. The country's future prosperity may someday be measured by computers and raw computer power per capita. Kierans argued that data processing expenditures had to increase substantially to meet the competitive threat posed by the Americans. The department was equally concerned with harnessing computer communications services with the achievement of national economic and social aims. A national computer utility could be the instrument of national policy.

A Canadian Computer/Communications Agency was proposed to coordinate the new utility. The DOC predicted that the computer services industry would become one of the three largest industries in the country in ten years. The state's interest in its growth came to be expressed in the desire to encourage the rational and efficient growth of national computer and communications systems, and to propel economic productivity in the process.

The Computer/Communications Agency had already existed in embryo within the DOC in the form of a task force on the relevant aspects of the question, and $5 million had already been earmarked for 1971–2 for the creation of a national computer communications network. The ambition of its promoters was to prevent total foreign domination of the computer and communications industries. More urgently, it seemed, the cabinet expressed the keen wish to ensure that computer centres and information be kept in Canadian hands and under Canadian law.

The proposals were grounded in the traditional concerns of public policy. By fastening telecommunications and other utilities to the common good, it was hoped that the state could exert some control over computer technology. The DOC wanted to assure that the advantages of computer power were properly disseminated throughout the country in the most cost-effective fashion.[27] Only then would it meet the state's social and economic development objectives. The newly printed Telecommission studies that had examined all aspects of telecommunications provided "graphic evidence" of the significance of this new utility to the future of Canada. And although many areas of uncertainty remained, Kierans and his officials argued forcefully that this was the chance for the state to put its imprint on the young industry.[28]

The government's best predictions about the growth of the computer services industry were truly astounding. Forecasts suggested that the computer services industry would "equal or surpass the auto industry" in size and contribution to GNP. It would also carry vital consequences for Canada's social and political structure, although those consequences were usually left undefined. A computer utility policy would have to define the field. The marriage of computers and communications would usher in an era of great changes which would be powered in Canada mainly by foreign-owned companies. Over 80 per cent of the domestic data processing market was supplied by companies outside Canada. Losing any control over what might become the largest industry in Canada and "the most important determinant of the future quality of Canadian life" would be a disaster of the first order. Another peril to be avoided would be the capture of the utilities by special interests that would favour one region or class over another. The very viability of Canada would come under question. On the other hand, the merger of computers and telecommunications held out the promise that the country would become "one great information system" with the computer as common as the telephone is today. The levels of creativity and leisure time unleashed would be added to Canada's economic and social health.[29]

The federal government did try to do something by underwriting 40 per cent of Control Data Canada's $56 million expansion program in the fall of 1970. CDC felt that an indigenous computer industry was a vital prerequisite to the industrial expansion of the country and the company aspired to become the backbone of a new Canadian computer industry. The finance minister, John Turner, was reluctant to commit to such a scheme, especially since any further federal support to CDC might affect the position of Canadian owned and operated firms. And significant Canadian equity in the company would be rejected by the parent corporation if it involved risking loss of control. The company also received over $6 million from Ottawa to develop the PL 50 computer.

The problem of supporting CDC was becoming a familiar one. In order to promote an indigenous computer industry, the federal government was compelled to subsidize an American multinational. Without that base, both sovereignty and the economy would be seriously weakened. The production performance and import statistics for computers and electronics for the 1970s shown on tables 24 and 25 demonstrate that the plan was not altogether successful. These results also highlighted some of the risks in forming a policy in the computer field.

As policy-makers were finding out, talking about a national computer/communications policy and acting on a plan were different things. Not only would the policy have to be assessed on the contribution it made to the achievement of government aims, but it would also have an impact on the traditional (and powerful) monopoly public communications services. What concerned policy-makers most was that any policy designed to achieve national objectives would meet with tough scrutiny by the provinces. Indeed, by 1970 some provincial carriers were planning to enter the data processing business, and Québec Telephone was already offering such services. The construction and operation of a national computer/communications system would also require the direct participation of the provinces which could result in conflict, especially considering federal plans for a cross-Canada computer network with heavy federal control.

Pressure to clarify the state's interests led to yet further DOC studies under the aegis of the Canadian Computer Communications Task Force. The task force chairman, Dr H.J. von Baeyer, was to direct a series of studies that would form the basis for a new technology policy.

Several months later, in May of 1972, von Baeyer presented a two-volume report entitled *Branching Out*. The task force emphasized the need for governments at all levels to recognize computer communication as a key area of social and industrial activity. Specifically, the report advocated maintaining and developing a competitive and innovative industrial environment, fostering the development and self-reliance of the industry, and ensuring a proper degree of Canadian independence. The report stressed that competition was the most effective vehicle for the development of the data processing industry and that the federal government should play a role as "focal point."

The concept of a computer utility, extensively discussed two years before in cabinet, was dismissed: low demand and a variety of services did not fit the utility model. Despite this, the idea of a national computer utility persisted for a few years; policy-makers were reluctant to relinquish the idea. What really killed it were suspicions that the "utility" in "Computer Utility" meant a radical change in the way business was done and more state interference.[30] By 1973, large service bureaux, among them the IBM Data Centres, offered computing power nation-wide (excepting the north) and smaller firms covered the rest. Lack of confidence (well placed, in retrospect) in the long-term survival of commercial systems operators that provided raw computer power was also an important factor in business attitudes. The lack of a clear jurisdiction in computer communications at any regulatory level, however, was one of the more obvious institutional problems. Government policy was developing into little more than the desire to respond to innovation and to keep data flows within Canada. The gulf was slowly widening between the stated importance of computer communications and concrete actions by the state.

Reaction to *Branching Out* was varied. The Consumers' Association of Canada argued that not only were the task force's terms of reference "grossly inadequate to the magnitude of the task," but the process was also highly institutionalized. The result was that the task force was "more responsive to the needs of Canadian institutions – business, government, universities – than to the needs of the Canadian people." The CAC lamented the emphasis on economic growth. With some justification, the association complained that the social impact of computer communications had become a "poor cousin and something of an embarrassment to the general developmental thrust of computer/communications policies." The Vanier Institute of the Family echoed the CAC's concerns and also bemoaned the lack of public input.[30]

Not surprisingly, Bell Canada and IBM Canada were in basic agreement with the recommendations contained in the report, the latter being careful to support "Canadian control, not Canadian ownership." Bell reminded the minister, however, that "planning and other managerial functions of industry [or] the legislated responsibilities and regulatory authorities "should not be tampered with. Canadian policy would not fundamentally alter the relationship between the state and the industry. Rather, it would attempt to increase the Canadian presence by stimulating the industry. This would mean strengthening the computer communications service industry and establishing policies to protect the Canadian market from an American industry that viewed the Canadian market as an extension of its domestic operations.[31]

The reaction to the proposals of the task force was essentially split along lines that emphasized size and function. For example, the data processing organization CADAPSO as well as the Electronic Industries

Association of Canada (EIAC) were strongly in favour of state-led Canadianization. EIAC strongly supported a "buy Canadian" policy, but wished that the government would stop engaging in bureaucratic shuffle and get down to policy implementation. However protectionist it was of the home market, EIAC warned the federal government to refrain from any measures that would prevent Canadian industry from having access to data processing services that would hamper its competitive status. "It is a dangerous fallacy," the EIAC brief to the DOC noted, "to believe that somehow a government can shield its service industries from international competition while asking is manufacturing sector to accommodate lower tariffs and to increase its export performance."[32]

The telecommunications common carriers had made their views abundantly clear to the federal government well prior to the release of the task force report. They were anxious to remind it that the managerial prerogatives they retained best served customer requirements. The Canadian Telecommunications Carriers Association noted that the "introduction of Dataroute by the Trans-Canada Telephone System, and Infodat by CN/CP was a worldwide first for Canada in the digital transmission field."[33]

Provincial responses to the task force report are indicative of how far the issue had been subsumed into the general conflict over communications. Ontario responded to the report by reminding the federal government that the basic issues raised could only be answered by policy and federal-provincial negotiation at the ministerial level. Quebec noted the lack of consultation with the provinces, and Saskatchewan would not discuss it unless the entire field of communications was discussed.[35] The disagreement over control, however, masked the essential agreement with the proposals of *Branching Out*, especially over the issue of restrictions on data processing, and the involvement of the common carriers in the field.

To be sure, Ontario's interest in computer communications was not always identical to other provinces. Those which had carriers were more liable to press for solutions that were based on the interests of their carriers. "We should be wary," warned one Ontario official, "and once again refrain from allowing other provinces to coerce us into a policy which we would find difficult to live with." Ontario was home to the majority of data processing firms in Canada. Given a choice between going along with the provinces on carrier entry and harming the industry, Ontario's interests should come first.[36] That interest in carrier entry was limited, but it stressed that no horizontal relationships and no discrimination exist. Most of all, if the financing of the data processing organization would have an effect on the rate of return, then it would invariably affect subscriber rates as well.

The problem of increasing the Canadian presence in the computer communications industry was akin to the problem that had dogged Canadian state and private enterprise for most of the twentieth century. Once a sector was effectively staked out by foreign capital, there was little state intervention would or could do. The import statistics for computers and electronic parts shown in table 26 are startling confirmation of the trend.

The federal government could restructure tariffs on data software and equipment in the procurement of its own data processing needs to meet its objectives. The Department of Industry, Trade, and Commerce would be a central player in any attempt to Canadianize the sector, but just what it could do was open to question. "The question arises," one DOC document suggested, "as to whether past experience is directly relevant to the present situation." One of the differences was that various government departments were attempting to coordinate their procurement policies to meet similar objectives. Perhaps the object of the stimulation – computer services – could also be better targeted than other service industries. Finally, computer communications was new and should thus be treated in a manner unlike anything preceding it.

As early as the spring of 1972, computer manufacturers and firms argued before the Tariff Board that a revised tariff structure was the solution to the problem. Data processing equipment and parts were subject to several tariffs rates, and several Canadian firms argued that this created barriers to growth in the industry. Canadian companies in particular were anxious to reduce the tariff on components, but not on end-product computers. Datagen Canada warned that if this continued, it would certainly cause the company to review its decision to manufacture in Canada. American branch plants protested too. The issues that the existing tariff structure created a "high level of confusion and harassment" because of its antiquity.

One of the principal recommendations issued by the task force was the establishment of a focal point to act as the locus for decision-making within the federal government. This could coordinate policies on computer communications issues as well as provide the government with a source of information on the industry.

For policy-makers, the issues were daunting: should they support competition, regulation, or national control? The prime aim in computer communications was to direct the technology to promote economic and social aims, emphasize national identity, and maximize

Canadian influence and control. There was also little agreement about who exactly would lead the government charge in this matter. The DOC felt it was the natural place for such discussions. The Department of Finance had different ideas. Finance believed that the Ministry of State for Science and Technology would have been a more appropriate place to achieve coordination, since it was established to perform such a role.[37] The dissent illustrates the difficulties within the government over the direction and control of computer industry development.

Apart from dealing with the carriers, higher-level policy considerations quickly presented themselves. The impact of the traditional monopoly in public communications had to be considered, as did the interests of the common carriers generally. Competition in the form of new carriers constructing new transmission facilities would be undesirable. In the long-haul communications field the DOC considered the effectiveness of competition limited by the nature of the service. Policy-makers believed that the integrity of the monopoly services must be safeguarded. In any event, extending competition would undoubtedly incite serious provincial opposition.[38]

Open competition could potentially involve further foreign penetration. The carriers considered any competition to come only at their expense. The benefits of competition were believed to increase with traffic density, but would have adverse effects upon the regional distribution of computer services. Additionally, the benefits accruing from a liberalized competition policy would accentuate the disparities between large and small firms. Of course, the large firms would be the principal beneficiaries of a more open competition policy, since they controlled over half of the market. Not surprisingly, the DOC concluded that "scale effects resulting from more effective loading of data transmission facilities will tend of work against Canadian ownership of computer services." The department concluded that the effects on competition, especially between the TCTS and CN/CP would not improve the Canadian computer services situation. The market could not meet the social and economic objectives of the state.

By 1973, another problem that had accompanied the rapid development of computer communications was beginning to emerge: the north-south flow of information. The DOC identified two key problems here. First, employment opportunities available in Canada were being eroded. The diversion of computer communications systems to the United States would lead to a serious loss of highly educated personnel and with them, technological sophistication. Second was the loss of national sovereignty over information. Foreign data banks storing Canadian data may affect the country's autonomy, and they would be beyond the reach of Canadian governments and courts.

Allan Gotlieb argued that data flows were contributing to a redefinition of sovereignty not along geographic or spatial terms, but along informational lines.[39]

Gotlieb and others argued that the physical location of data banks was important in regulating access to them, and in the state's ability to regulate them. The flow was predominantly and predictably north-south, which might result in a "national self-perception of impotence, an ability to effect one's vital choices, and the effective erosion of one's political sovereignty." Computational power, like communications and electric power, was regarded as crucial to the survival of the country.

At the same time, Canadian firms were already dependent on American services. Restrictions on international data flow would create difficulties for companies such as Canadian Pacific and Air Canada. Canadian firms were anxious for government control only to prevent unfair competition in predatory or discriminatory pricing, tied sales, and exclusive dealing or refusal to deal.[40]

One of the areas where the state could direct the course of computer communications was in education. Canadian universities were beginning to purchase computer services via networks linked from American institutions. The implications were evident: not only would computer communications' Canadian content decrease, but it would also allow the compromising of research and development objectives as well as cultural activities.

Canada's proximity to the nation that could boast the most advanced, sophisticated computer communications tools placed Canadian policy-makers in a serious quandary. The goals of patriation of economic resources met squarely with the desire to have the most advanced US tools to aid in management and service processing, and the reality of a 25 to 30 per cent duty on computer equipment in Canada. The situation needed attention. On the one hand, the clear objectives of Canadianization needed, at least for a time, strong fiscal support; on the other hand, the consumers wanted cheap and state-of-the-art technology. The uncertainty about government policy in the computer communications field was viewed as a serious obstacle to the development of the Canadian industry.

Trade in services flourished, mainly provided by US-based organizations selling services to Canadian companies. In-house services which were provided by multinational corporations to their branch plants in Canada also accounted for a significant proportion of the traffic. The situation led right to a straitjacket: higher tariffs might protect some of the Canadian firms but they would be costly for Canadian firms, which would continue to import data and software, only at higher prices. A hike in the tariff would lead to serious consequences for the very industry it would have been designed to protect. Any policy would run the risk of becoming a dead letter because of the difficulties involved in differentiating physical media (such as disks and computer printouts) and imports by electronic transmission.

Lack of access for Canadian computer service companies in foreign markets made the situation worse. The task force predicted that while in-house data processing would not become a problem of international competition, the field of public computer communications and computer-based information services would. The anaemic performance of the Canadian computer services industry would be a testament to this, as table 27 demonstrates.

The government had to reconcile the irreconcilable; it had to protect the country's economic and social fabric, while preserving private initiative and innovation. Cabinet ministers assured that new computer communications technologies would not be allowed to override the older telecommunications configuration.[41] Even within the federal government, Treasury Board officials were wary that the proposals would commit the government to a restrictive course of action. Specifically, they were concerned that Industry, Trade, and Commerce square the recommendations with their previous suggestion to the multinationals operating in the computer industry that they set aside a proportion of production and R&D for development in Canada.[42]

By 1974, Ottawa's attempts to develop a national computer communications policy made three things clear. First, the complexity of the technology was matched by the diversity of interests at stake in any federal initiative. Secondly, the problems of computer communications did not attract the political interest it once generated. Third, the centrepiece of federal initiatives – the computer utility – did not have much hope of achieving any consensus, especially when provincial ambitions were beginning to emerge in the field.

PROVINCIAL RESPONSES
TO CONVERGENCE

If provincial interest in the field did not parallel federal attention, it certainly mimicked it. If any independent computer utilities sought to establish a national network, federal power could be invoked. Still, federal jurisdiction was by no means certain, and divided jurisdiction could hamper Ottawa's plans to introduce a national computer communications policy. How were the provinces responding to these technological developments?

Although provincial governments had no special responsibility for the computer industry in general, by the early 1970s computer communications had nonetheless attracted significant attention. The link with communications had triggered an even closer provincial interest, and for good reason. Some provinces still retained primary responsibility for telecommunications regulation; others wished to connect their province to the promise of the new technologies.

Partly as a result of the discord over computer communications, federal-provincial relations had been taking on increasingly confrontational dimensions. Behind several important carriers stood provincial governments with communications aspirations themselves. It was thus doubly important to understand the position of the provinces. The big question was whether the provinces or provincial communications carriers would adopt the suggestions of the federal government contained in the task force recommendations. The federal government could find itself in support of CN/CP against provincial authorities defending the interests of their regional carriers.

The Saskatchewan response to the proposals was indicative. SaskTel responded bluntly to Ottawa's proposals over increased competition: "Noneconomic social objectives for an industry require a utility ... equitable access requires a utility ... 'no duplication' requires a utility."[43]

In the spring of 1970, the issue of computer communications had attracted the concern of policy-makers in Ontario's Ministry of Transportation and Communications. The computer utility generated the most early interest in Ontario. It was not long before discussion turned to parallels with Ontario Hydro. James Shantora, chairman of the Telecommunications Committee formed in 1970 to consider Ontario telecommunications policy, put the two together:

In discussing computer networks an analogy might be made between the computer utility and the present hydro electric utility where different generating plans in the province of Ontario are tied in with lines, not only in Ontario, but are connected into a hydro bridge system that stretches all the way to Florida and to British Columbia through the northern United States. In thinking of a computer network some standardization is required as to language, common way to represent characters, etc. ... The analogy can here be drawn between the hydro electric networks where the frequency and the voltage and the electrical outlets have to be standardised before any type of grid system could be established.[44]

Although the committee came to the conclusion that a computer utility was not needed immediately, a couple of trends were apparent.

First, its significance lay in the relationship to the telecommunications field as a whole. Yet no government, federal or provincial, had staked a claim. Second, somebody had to do something: one government or another had to guide and coordinate the industry. As Shantora argued, Canada needed "a right environment *now* in order that a computer utility can evolve and grow, in an orderly fashion."[45]

The committee made several recommendations to the MTC. It advised Ontario to move quickly to establish a regulatory body that would have authority over data classification and exchange. Failure to do so would mean leaving a vacuum for the federal government to fill. Ottawa planned to occupy the entire field, and Ontario's interests would left to the sidelines.

The establishment of a computer utility raised another question: Where would the money come from? Ontario feared that Bell would be the chosen candidate. If the utility's costs were underwritten by telephone subscribers in Ontario and Quebec, those provinces would be carrying an intolerable financial burden. Added to that was the problem of the standards of the small independent telephone companies which often could not measure up to urban telecommunications demands. This was a powerful stimulus for the province to "obtain a 'piece of the action'" and ensure that both governments and subscribers were protected against arbitrary plans by Ottawa.

Ontario had its own interests at stake in any new computer communications network. The provincial government was one of the largest users of computer communications in the country and was concerned that federal interference might damage its plans for expansion. The MTC minister, Charlie MacNaughton, was sure to make his government's concerns to Gérard Pélletier by stressing his concern about any economic controls over an area Ontario wanted to develop.[46] The MTC also felt that Ottawa had failed to take into account regional needs in the execution of national policies. That process had been costly to Ontario "as the province has been required to remedy regional deficiencies." The method by which Ontario wished to approach the matter was not based on constitutional powers, but effective "practical" concerns. This meant negotiating with Ottawa for shared control over a possible Canadian computer communications network, and close consultation before any regulation was put in place.

As early as the spring of 1971, Premier William Davis argued that "If plans for national data transmission systems which will become ever more central to economic development are not framed in a manner consistent with Ontario's own economic priorities, this province will be denied an indispensable development tool."[47]

The issue that the province had identified as important to economic development was mainly the problem of foreign ownership. Since the bulk of Canadian commercial data processing was done in Ontario, it was natural that the province should press for a more forceful and nationally oriented policy of foreign ownership and control. The threat of foreign penetration into the data processing industry led the province to throw its support behind the entry of common carriers into the data processing field. The support was granted with a condition: that the corporate and regulatory environment be structured to prevent unfair competition and cross-subsidization.

Ontario was also concerned about the capital required needed to keep the computer industry viable. "There will be revenues accruing to government from a highly developed computer communications industry," one official reminded his colleagues, "granted insistence upon Canadian ownership and the province's skill in obtaining its fair share of such industries."[48]

Quebec's position in "téléinformatique" was linked closely with its general position on communications – that full provincial autonomy should prevail in the field. The real fear in the computer communications field was that Quebec would be left behind other regions in the development of this new technology. To avoid that possibility, the Ministère des Communications du Québec (MCQ) established an agency in September 1973 that would deal exclusively with computer communications policy and problems.[49]

The Comité Interministeriel sur la Téléinformatique au Québec (CIT) was charged to develop the essential elements of a computer communications policy for Quebec, and work out coordination strategies for Quebec business and government. The work of the CIT went a long way towards articulating the province's position. The message was simple: a "patrimoine téléinformatique" was essential for economic development and cultural viability. Otherwise, technology menaced Quebec's very cultural and economic survival.[50] Quebec computer communications policy also reflected other traditional concerns such as economic and regional development.

The principal motivation for Quebec's entry into the computer communications field was defensive. The communications minister, Jean-Paul L'Allier, was wary of "un certain colonialisme de l'informatique" from other parts of Canada and the United States. The plan was centred around having the computer communications field in Quebec occupied by a Québecois institution, thereby ensuring effective provincial control over the development of the technology. The problem was located in Ottawa. As an MCQ official wrote to the deputy minister: "Cette politique pourrait demeurer sterile cependant s'il le fédéral conserve l'initiative du développement de la téléinformatique, initative qu'il cherche à consolider part le biais du control qu'il exerce sur les principes sociétés de télétransmission de données ainsi que sur les banques à charte."

For its part, Quebec viewed the computer communications field as part of the range of services that should properly be administered at the provincial level. L'Allier was particularly concerned with keeping Quebec's competitive position on the international market. Comparing its importance with transportation, he stressed that the "autoroutes electroniques qui pourront transmettre des donnés, des images et des sons," would be just the beginning. Echoing his stress in other areas of communication, he added that "nous voulons en outre éviter que d'autres decident pour nous – lorsqu'util pour nous – si'il vaux mieux participer à tel ou tel centre plutot qu'à tel ou tel autre centre."[51]

By the late 1960s, the Saskatchewan government was anxious to keep abreast of technological trends in computers and communications. As Gordon Grant reminded his audience at the opening of the Saskatchewan Government Computer Centre in 1968, "I think all of us can take pride that the Saskatchewan Government and its agencies is [sic] in the front ranks in using this new technology."[52] The real push for the installation of computer technology was the establishment of Saskatchewan's Medical Care Plan. A government in search of a cost-effective health care system was quickly sold on the idea of saving $243,000 per year on its record system using the computer instead of conventional punch-card equipment. The IBM Model 1410 was put into use, administered by the Government Computer Centre.

Acting on their evaluation of the provincial market, and anxious to capitalize on the increasing demand for computer processing, the Saskatchewan government established the Saskatchewan Computer Utility Corporation in 1974. The corporation was severely limited, however, by the lack of market formation and the "lack of history" of the computer industry in the province.[53]

The Corporation was established primarily to stimulate economic growth by supplying computing services to provincially based enterprises. A large proportion of its sales would come from inside government as well. The Corporation grew slowly, but seemed to adequately service the provincial need. At the end of 1974 there were twenty remote installations, compared to seven the year before. The total operating budget was just under $4 million in 1974.

The interests of all the Canadian provinces in communications were similar enough to promote series of alliances between 1971 and 1972. Concerns about the federal government's decision to use CN/CP Telecommunications and Telesat as second and third national carriers elicited concern and consternation in the provincial capitals, especially those who had their own carriers.

The complaints from western communications ministers were that "federal initiatives in the field of voice and data communications, cable communications ... computer communications, educational communications and remote-area communications among others, do not take account of provincial policies, priorities and jurisdiction."[54] Computer communications in particular should come under the provincial control either through legislation, carrier entry, regulation, or even through a public utility model under government ownership and control. Western Canadian communications ministers were particularly anxious to avoid unnecessary duplications of existing utilities, and increased rates.

Several provinces responded to federal actions in the computer communications sector by establishing their own initiatives in the field. Provincial interests in computer communications were eventually combined into the larger conflict over communications policy. Provincial actions served to further stall Ottawa's already troubled plans for a comprehensive national computer communications strategy.

CONFUSION AND FORFEIT

Provincial initiatives in the computer communications field merely added to the federal government's own confusion about the direction of its policy. By the mid-1970s, policy-makers had to contend with two major problems: government ambitions had to take into account the divergent interests of corporate players suspicious of federal motives; and the state did not speak with one voice on issues of computer communications policy, making a consolidated effort difficult. As a result of the centripetal forces at work, by 1975 the federal government abandoned its attempts to fashion a comprehensive policy in the sector.

Disagreements over computer communications and a national computer utility had a specific effect. A concerted policy was stalled, allowing the larger companies occupying the field the maximum freedom of movement. What is more, the federal state's reluctance was fed by a dim awareness that computer communications services would not evolve in the utility direction. One expert expressed it in the following terms: the data processing industry, rather than becoming a utility, will become more like the retail food industry, realizing economies of scale for mass products through automated agricultural and supermarket networks, but also providing a rich variety of speciality food and services, ranging from gourmet and diet stores to 24-hour delicatessens at premium prices and dockside fish markets at minimum prices.[55]

By 1974, it was evident that the question of what organizations should be permitted to provide which services in the computer communications field had been resolved in favour of a more competitive environment. But the quandary of competition versus Canadian control still remained. Competition would be maintained only at a significant cost if large financial institutions such as the common carriers or the banks were permitted to utilize their extensive resources. That situation would favour Canadian control but be underwritten by well-financed Canadian corporations.

The idea of a computer utility had been seriously questioned by the task force studying computer communications. Despite the recommendations of *Branching Out*, policy-makers in both Ottawa and some of the provinces still viewed the development of large computer communications systems as a golden opportunity for the development of national and regional computing capabilities. Raw computing power would be delivered to thousands of consumption points along an informational power grid. The potential to reduce regional disparities and staunch the north-south flow of information were cited as attractive features of the plan.[56]

The computer utility would provide services such as system design, programming consulting, data processing and facility management – all on a nationwide basis. Economies of scale would drive the costs of data processing down, providing a technological boost to Canadian firms. If the idea of a computer utility found resonance with policy-makers in Ottawa, the private sector was more suspicious of the idea. As we have seen, many reacted negatively to the very term, especially given the lack of familiarity with the new technology most businessmen possessed. Some were convinced that the term described a covert intent by the state to change the environment of computer operations from a competitive, unregulated field into a franchised monopoly service such as other utility sector undertakings.

The difficulties of policy-making in computer communications led some within the government to call for increased public participation in the process. The Secretariat of State in particular lamented the limited nature of public input into computer communications policy. "Currently in Canada," it argued, "policy is in effect formed by the government and the corporations involved in computer communications, both producers and consumers of such services."[57] The state had a positive obligation to guide the technology in the public interest, and was not doing so.

Other government departments were also keenly interested in the démarche of the new technology from their particular perspectives.

What emerges is a cacophony of concerns and claims to have the government harness the new technology to a range of imperatives. The Department of Regional Economic Expansion argued that a properly configured computer communications policy could cement national unity "by supporting the kind of feedback and the opportunity for interaction between socio-cultural groups." The idea was amplified by the suggestion of Manitoba's Communications Minister Ian Turnbull, who proposed that the federal government decentralize its computer installations to less advantaged areas of the country as a tool of development.[58]

But by 1974, the general principles of Ottawa's computer communications policy had been set. The state was to act as a catalyst, fostering policies and financial measures which would create the right kind of environment for some degree of Canadian ownership and control, not to mention regional equality.[59] It was careful to emphasize that in the data communications field it would not "undermine economies of scale" and cross-subsidization that was the basis for regulation of the public utility. There were the usual commonplaces about the transformative nature of the technology, and the importance for national integrity and international competitiveness contained within it. The reliance would be on competition policy to ensure Canadian consumer demand was met. To that end, the government elected to abstain from direct regulation of the industry. In the course of five years, the state had been largely unable to mount an effective defence of its own interests, however modest.

In January 1975 the federal government announced policy guidelines relating to the participation of federally regulated carriers and the chartered banks in data processing. The policy reflected the twin objectives of maintaining a competitive market structure and the promotion of a strong Canadian presence in the provision of computer services. Ottawa supported the concept of a nationwide common user communications network, but made no attempt to identify the role of the two carrier systems, TCTS and CN/CP. The announcement of the policy was not a major departure from what had been discussed within the policy groups in the previous two years.[60] In the meantime, discussion about the implications of computer communications had been going on in the Computer/Communications Secretariat in the DOC since mid-1973 through a series of committees. The deliberations resulted in a series of final reports in mid-1975.

The lack of vigorous government action had satisfied some elements of the computer industry and exasperated others. The Canadian computer services industry had long been a topic of discussion within the federal government. It had also been perceived as a key industry. But by 1975, little but good intentions had been expressed towards the industry.

The computer service industry's main source of apprehension had been the activities of computer manufacturers in the business of computer services. The opportunity to publicize their discomfiture came at the hearings of the Royal Commission on Corporate Concentration in September 1975.

As president of Systems Dimensions Limited, George Fierheller had cause to worry. His company was the major independent supplier in the computer services business. He told the royal commission that foreign-owned computer manufacturing firms were pressuring Canadian business out of the market. The computer manufacturers, especially IBM, often integrated their computer services business with their mainframe operations, rendering it impossible to assess the true cost of the services that were offered. Although the pricing problem may appear insignificant to the multinational manufacturer, Fierheller argued that it created chaos in the domestically controlled market.[61]

Fierheller asked for the implementation of the government's own report on the problems set out in *Branching Out*. The domestic industry wished to have the manufacturers establish an arms-length subsidiary, much as the telecommunications carriers were required to do in their cross over to data processing. The argument found resonance within the DOC, but little could now be done.

Jump-starting the Canadian computer communications industry continued to be a concern, and for good reason. In 1971 over half of Canadian industry used computers. "Like transportation and energy," Douglas Parkhill wrote to Deputy Minister Max Yalden, "the products and services of computer/ communications play in indispensable role in the functioning of many other industries."[62] Competition since then in the provision of services had driven General Electric and RCA out of the field. The independent companies that had emerged in Canada by the late 1960s capitalized on the trend away from in-house computing. By the mid-1970s, however, the expansion of the original concept of the computer utility had changed. Policy-makers believed that this concept would result in larger and more sophisticated utilities.

The trend would offer a stabilizing effect on the economy, since firms could minimize cash flow problems by avoiding large in-house capital costs. Process control systems, airline passenger reservation systems, payroll and large data bases could all be harnessed to a large computer utility with relative ease.

Here was the worry to those who sought to retain a semblance of Canadian influence in the industry. The United States was well advanced in the field, creating wide discrepancies between American and Canadian capabilities. This problem knew no boundaries, given the ease with which services were offered across the public switched telecommunications network. If Canada was unwilling to submit to foreign control in essential services, appropriate measures would have to be taken.

The need to create large Canadian utilities would be costly; it was estimated that the market would have to be over $500 million just to break even. The critical size for the Canadian utility would be a market volume of around $10 million, a level that would enable a Canadian firm to compete against the more powerful southern utilities. The options open to Ottawa were the ones that had always existed: create or support a national champion, or support multiple ownership of shared facilities. Increasing reliance on the carriers seemed to be inevitable in any case.

CONCLUSION

By 1975, computer communications had been installed onto the hard drives of Canadian communications. That installation first generated ambitious plans, but aspirations for public control failed, and miserably, too.

This outcome was not surprising. Large corporate interests prevailed over smaller concerns, particularly of the domestic computer service industry. A lack of policy guaranteed the position of the well-established enterprises, at the expense of the newcomers. Furthermore, advocates of a vigorous policy in computer communications could not mobilize authority within the state to the cause of directing this technological transformation. That blockage weakened the state's attempts to put its imprint on this technological innovation. Finally, federal-provincial conflict stalled the promotion of federal policy in this arena.

The failure of the Canadian state to develop a policy had consequences. First, hopes that national and public priorities would figure prominently in computer communications were frustrated. In spite of the aspirations of the more determined policy-makers, national policy had confronted the new technology with a mixture of naiveté, ambition, confusion, then resignation. The result was that Canadian priorities were subordinated to other concerns.

Secondly, the attempts to frame the historical model drawn from the state's experience with utilities surrendered to the pressures and experience of the day. Even if the technology could support something approaching a national computer utility, the political and economic impetus crucial to its implementation existed only in traces.

Instead, another model imposed itself upon the new technology. In place of a utilities development model, the nature of federal involvement in the computer communications field reflected the compromise reached between state and corporate power as well as between federal and provincial interests. The Canadian computer services industry would have to work the boundaries of that compromise. The hope of 1969 had surrendered to the regrettable facts of 1975. It need not have been so.

Third, the idea of computer communications as a common property like telephony and electricity was suppressed in favour of letting the market decide. Market distribution of computers and computer communications accentuated regional disparity, in which Toronto and Montreal were the principal beneficiaries. Economically marginal provinces such as Saskatchewan relied on the state instead to distribute technological blessings. The new technology was diffused in a remarkably rapid fashion across the commanding heights of the Canadian economy, but without the application of utility-based principles.

The Ontario case was fairly representative of provincial initiatives in computer communications. By 1976, the MTC had decided that a crucial decision point on the issue had been reached. But Ontario's policy, such as it was, had not borne any fruit. Ontario had made major commitments to develop a policy, but its resources were limited. The courses of action were also limited. Queen's Park had assigned the issue of computer communications a lower and lower priority in the communications field. Policy discussion then revolved around damage control. As one official acerbically wrote, the government could "fake it and tart the subject at a superficial level and attempt to present a plausible appearance to the world."[63] The other alternatives were either to vacate the field or recommit personnel and "build from the ashes." Amateur meddling in the industry would do more harm than good; the government would have to commit itself to coming up with a policy. As it stood, however, policy-makers were content to let the private sector decide the parameters of the possible in computer communications, and rely on input to the federal government to influence policy.

The irony about Ontario's foray into the computer communications field is that the rhetoric about the computer revolution was swallowed, but very little was done about it. By 1976, the Communications Branch could point to only three concrete achievements: liaison with the federal government, which it conceded had seized the initiative in this area; some policy discussions; and a plan to recruit and train a suitable computer scientist for permanent staff.[64]

The Canadian experience with this technology provides an interesting insight into the way governments viewed the introduction of innovation. The technologies were sanctioned by their revolutionary potential to transform economic and political relationships for the better. Intentionally or not, the advertising was misleading: the revolution did not happen. The technological transfiguration of society would happen in a diffuse manner, not in a highly focused way.

The state's experience in the field ultimately revealed its ambivalence in the face of the convergence of computers and communications. On the one hand, it harboured deep suspicions over whether the market could deliver on the national interest. On the other hand, advocates of reform could not mobilize the full authority of the state to effect serious reforms. The consequence was disjointed policy and an inept choice of tools. In the process, it exposed the limits of the state's capacity to act.

Both national and provincial policy-makers had choices to make about how to direct computer communications. The role of technology was important, but not determining. The compromises reached reflected the demands of the private users of the technology, corporate and continental interests, and only partially, the interests of the state. The asymmetrical relationship of those interests asserted themselves to the detriment of the national and public interests as they had defined it. Canada's confrontation with computers had altered the state by weakening its future potential to stake a claim against continental or technological imperatives.

Conclusion

In the first two decades of the twentieth century, the struggle to establish a balance between the interests of both producers and consumers in telecommunications led to the creation of a regulatory regime. The socio-economic conflict created by monopoly control of a key public utility was solved. The agency charged with balancing those interests, the Board of Railway Commissioners, restored a political and economic stability to the operation of telecommunications. The organizational and regulatory equilibrium thus created lasted into the postwar period.

The period between 1945 and 1975 witnessed a number of changes to the political, economic, and social context. Canada was enmeshed in a series of international arrangements and alliances in which it had to define and articulate its interests as a middle power. Economically, a period of prosperity and a rising standard of living were mitigated by a skewed pattern of income distribution and a capital concentration under private control. In the social sphere, the values of consumption took hold, and gave birth to a movement dedicated to protection of the consumer. Both the diffusion of the telecommunications network and the introduction of new techniques would take place in this milieu.

The challenge to public policy in this environment was multifaceted. Within the context of a liberal democratic and federal state, Canadian policy-makers in telecommunications had to safeguard national sovereignty in the international arena. At home, the challenge was to ensure that the regulatory instruments of the state provided effective mediation between the utilities and their subscribers, and distributed burdens with a degree of fairness. The government was also confronted with the task of ensuring that the economic organization of new technologies involved a degree of public choice, whether this meant paying heed to securing national unity, or disseminating the benefits across society equitably.

The challenge to regulation was to maintain that balance in the face of the remarkable expansion, technological change, and consumer pressure to keep rates down, especially in times of inflation. Implicitly, federal regulation of telecommunications would also be required to adjudicate who was to participate in this expansion of the network, as well as who was to pay for it.

Federal regulation succeeded in providing a fair return on investment and safeguarding the industrial organization of the sector, most notably with the preservation of the relationship between Bell Telephone and its supplier, Northern Electric. Not least, the Board of Transport Commissioners provided an environment in which the telecommunications network flourished, providing tangible economic benefits for the utility and its subscribers.

Across the period of this study, Bell flourished. As tables 28 and 29 show, between 1945 and 1975 assets grew by an average annual rate of over 11 per cent, operating revenues by over 12.5 per cent, and net income by almost 14 per cent. The company was also able to maintain consistent per share dividends throughout the period, total dividends growing by approximately 11.2 per cent on average per year.

By international standards, the Canadian telecommunications network was indeed a success. As Tables 30, 31, and figure 1 show, Canadians were near the top of the league tables of industrialized countries when it came to all the standard measures of high achievement in telecoms: a high volume of national traffic, lots of telephones upon which to conduct those conversations or exchange data, and a high proportion of per capita income derived from telecommunications services.

To a certain extent, government regulation could share in that success. Or could it? Certainly, government policy could have prevented an even greater statistical performance. Conversely, it could have balanced economic efficiency with the public interest. Regulation's main weakness in the postwar period was its inability to adapt to new circumstances. Technological advance and the growth of the network generated demands for greater public input, but the Board of Transport Commissioners was unable both in form and procedure to provide a platform for the concerns of consumers. By largely standing aside and allowing the utility to dominate the thrust of development, the BTC was estranged from the shaping of telecommunications, and from those outside the regulatory arena who were affected by its decisions.

In telecommunications policy, a different but complementary set of considerations demanded the attention of policy-makers. In both domestic and international spheres, public policy had to grapple with two main challenges. First, government had to ensure that international arrangements, and specifically arrangements with the Americans in communications, did not limit the economic potential of the domestic telecommunications industry. Secondly, the state had to ensure that technological transformation such as satellite development took shape in a manner consistent with the imperative of maintaining national unity and within the boundaries prescribed by the federal structure.

What was the result? In the two decades after 1945, the federal state succeeded in defending its communications priorities in both the international and domestic political arenas. In concert with members of the Commonwealth, Canada agreed to subordinate market forces to political ones by operating its own overseas telecommunications links, despite suspicions about the efficiency of such a move. When Commonwealth ties proved less effective in the defence of the country's interests, Canadian foreign policy embraced international instruments as a counterweight to the preponderant influence of the United States over developments in telecommunications technology. In domestic politics, the federal state manœuvred to pre-empt Quebec's halting attempts to participate in a satellite scheme with France by moving quickly to launch a federally-sponsored domestic communications satellite. By the late 1960s, however, events began to overtake the ability of policy-makers to somehow keep up with the evolution of the technological and economic organization of the field. The issues generated by postwar telecommunications developments revealed the increasing inadequacy of Ottawa's administrative arrangements to deal with an important sector of economic life.

In telecom policy, as in regulation, the technological changes of the 1960s proved to be a stiff challenge to public policy. Policy-makers within the federal government viewed the impending communications revolution as an opportunity to revitalize a sclerotic democratic system, cement national unity, and place domestic industry on a solid footing. The impulse to rationalize the administrative capacities of the state to meet those challenges resulted in a series of bold federal initiatives designed to create a national telecommunications policy.

National policy-makers were quickly confronted with the federal nature of the Canadian polity. National plans were quickly matched by equally ambitious provincial plans, especially from Quebec. Federal attempts to promulgate a national telecommunications policy eventually foundered on the shoals of federal-provincial conflict. Federal insistence on jurisdiction and control in telecommunications were matched by equally insistent provincial demands. The predictable result was federal-provincial deadlock.

Compounding the challenge of federal action in telecommunications was the keen desire of established companies, most notably Bell Canada, to preserve the status quo as long as possible to assure market advantage when the march of events would compel changes in the institutional structure of the telecommunications industry.

A similar outcome resulted over computer communications problems. Federal action in that field was viewed as a critical instrument in asserting control over a largely foreign-based technology. But the same dynamic that forestalled comprehensive reforms in telecommunications regulation was hard at work in the computer communications field. The combination of industrial and provincial pressures and a weak federal political will made a comprehensive national policy impossible.

In the area of policy, then, both in telecommunications and in the field of computer communications, the promotion of a grand national policy failed. Federal-provincial conflict and opposition from the industry's major players resulted in that outcome, in spite of the best efforts of policy-makers. By 1975, failure to act resulted in a forfeiture of leadership for the federal government. Ottawa was left to exercise its control over telecommunications in a piecemeal fashion; in computer communications, big plans were replaced by modest ones.

The lack of agreement on telecommunications policy did not cause telephones to stop ringing, or modems to stop connecting computers. The pattern of industrial organization continued much as it had been, influenced by technological change and market forces. Instead, the lack of a comprehensive telecommunications policy reduced the chances that federal policy-makers could impose their priorities upon telecommunications matters except in an ad hoc fashion, through either cabinet intervention or by depending upon a sometimes unpredictable regulatory agency. Ottawa did not retreat from making telecommunications policy, but it was unable to produce an arrangement which would aspire to organize the entire sweep of telecommunications. Then, as now, opinion will be divided as to the desirability of having had such a well-articulated policy. For proponents of state intervention, the policy was a mitigated failure; for those who wished to have governments set the rules, but not participate in the game, this outcome was perfectly satisfactory. What seems evident, however, is that a tightly focused and strategic vision for government policy in telecommunications would undoubtedly have been able to accommodate more than it actually did.

The stakes in telecommunications regulation were much higher, since the consequences of not solving the problems of regulation could have much more immediate effect. Canadian Transport Commission decisions provided some measure of balance in a difficult environment. The commission did not cause telephones to go dead; making them ring, however, might be more expensive. Regulatory problems could also result in longer waits for new or upgraded services, delays in capital projects or industry layoffs, depending upon the circumstances.

Because of their importance, regulatory actions caused an opposite, but not necessarily equal, reaction. The provinces of Ontario and Quebec and restive consumer groups all demanded greater input into the regulatory process. An inflationary economy put further political pressure upon the CTC to hold costs down while Bell continually demanded protection for its operations. The commission's inability to handle conflicting demands of the utility, the state, and consumer groups over prices, technological progress, and the pattern of diffusion of telecommunications led to agency paralysis. The politicization of telecommunications issues in the mid-1970s further impaired the regulator's ability to adapt to changing circumstances. By the time the CTC transferred its responsibility for telecommunications to the CRTC, the regulatory tribunal had become the site of dissonance and conflict that transcended the boundaries of regulation.

Whether in regulation or telecom policy in Canada, solutions to new problems were often conditioned by traditional Canadian thinking about utilities and technology. By 1975, this resulted in a system in crisis – a system longer able to provide effective mediation between consumer demands and the concerns of the utility.

The federal and provincial governments examined in this study shared the conviction that policy control was the paramount issue in telecommunications. When these convictions were played out in a federal structure in the form of discussions about the establishment of a national telecommunications policy, the result was dispute and deadlock. By adhering to the reflections of past transformations, policy-makers pre-empted more pragmatic solutions to managing telecommunications and new technologies. In the process, the aspirations of those who wished to see a greater role – or at least a strategic role – for the state were frustrated.

For Canadian policy-makers, 1975 represented an end of sorts. That year, the end of CTC regulation of telecommunications, and the failure of a series of federal-provincial conferences on the subject, would indeed be the beginning of a new phase in the development of telecommunications policy and regulation in Canada. Subsequent attempts at fashioning a policy would rest on the legacy created by the state's previous management of telecommunications. The Canadian Radio-Television and Telecommunications Commission would embark upon a period of regulatory activism in the late 1970s, attempting to exert state power over the sector. In the nearly quarter century of CRTC regulation of telecoms after 1976, massive technological and market pressures have utterly transformed the once-ordered world of communications. A more open architecture in the sector has been the outcome of these changes, hastened along by market forces, regulatory decisions, and perhaps most crucially, remarkable advances in communications technology. Competition in the long-distance market, and soon even in the provision of local service, has shadowed trends in the United States. The state has been largely confined to setting the general rules of the game.

As this study has shown, in the period when these new technologies were just taking shape, both businessmen and politicians had a role in diffusing the technology. Telecommunications matters forced governments to define their priorities in an increasingly complex field. But the federal government was only partially able to press its claims upon the development of telecommunications. The legacy of those three postwar decades would have significant consequences for both the ambit of government policy in telecommunications, and over the Canadian state's future claims to leadership over technological change in communications.

Appendix: Tables

Table 1

Long-distance rates between Montreal and selected cities, 1918–1975
(rates for station-to-station, daytime, 3-minute call, in dollars)

Date			Between Montreal and			
	Ottawa	Toronto	Windsor	Halifax	Winnipeg	Vancouver
20 November 1918	0.70	2.05	3.25	3.20	7.10	14.60
21 April 1921	0.70	2.05	3.25	3.20	7.05	14.50
1 February 1929	0.70	1.95	3.05	2.05	4.00	8.00
1 January 1930	0.65	1.90	2.85	2.05	4.00	8.00
1 September 1936	0.65	1.90	2.85	1.80	3.75	6.75
1 April 1937	0.65	1.75	2.30	1.80	3.75	6.75
10 July 1941	0.65	1.75	2.30	1.75	3.25	5.75
1 September 1945	0.65	1.75	2.30	1.75	3.00	4.70
22 July 1950	0.80	1.75	2.30	1.75	3.00	4.70
15 May 1953	0.80	1.75	2.30	2.00	2.95	4.40
1 December 1958	0.80	1.70	2.10	2.00	2.95	4.40
1 June 1959	0.80	1.70	2.10	2.00	2.85	3.65
6 November 1960	0.80	1.70	2.10	2.00	2.75	3.50
1 July 1965	0.80	1.30	1.65	1.75	2.35	3.65
1 July 1970	0.80	1.30	1.50	1.75	2.35	3.00
1 July 1975	0.93	1.38	1.58	2.10	2.67	3.15

Source: Series Statistics Canada, Historical Statistics, Series T336-341.

Table 2
Long-Distance Rates between Montreal and Selected Cities 1918 to 1975, in 1981 dollars (rates per station-to-station, daytime, 3-minute call)

Date	Between Montreal and						Canadian Average	Decrease %
	Ottawa	Toronto	Windsor	Halifax	Winnipeg	Vancouver		
20 November 1918	3.89	11.39	18.06	17.78	39.44	81.11	28.61	n/a
21 April 1921	3.47	10.15	16.09	15.84	34.90	71.78	25.37	−11
1 February 1929	3.76	10.48	16.40	11.02	21.51	43.01	17.70	−30
1 January 1930	3.53	10.33	15.49	11.14	21.74	43.48	17.62	0
1 September 1936	4.33	12.67	19.00	12.00	25.00	45.00	19.67	12
1 April 1937	4.22	11.36	14.94	11.69	24.35	43.83	18.40	−6
10 July 1941	3.82	10.29	13.53	10.29	19.12	33.82	15.15	−18
1 September 1945	3.53	9.51	12.50	9.51	16.30	25.54	12.82	−15
22 July 1950	3.17	6.94	9.13	6.94	11.90	18.65	9.46	−26
15 May 1953	2.83	6.18	8.13	7.07	10.42	15.55	8.36	−12
1 December 1958	2.61	5.56	6.86	6.54	9.64	14.38	7.60	−9
1 June 1959	2.58	5.48	6.77	6.45	9.19	11.77	7.04	−7
6 November 1960	2.55	5.41	6.69	6.37	8.76	11.15	6.82	−3
1 July 1965	2.35	3.82	4.85	5.15	6.91	10.74	5.64	−17
1 July 1970	1.95	3.17	3.66	4.27	5.73	7.32	4.35	−23
1 July 1975	1.59	2.36	2.70	3.59	4.56	5.38	3.36	−23

Source: Compiled from M.C. Urquhart and K.A.H. Buckley, eds., Historical Statistics of Canada (Ottawa: Statistics Canada, 1983).

Table 3
Telecommunications in Canada: Telephones and Telephone Calls, 1946–1975

	Number of Telephones at 31 December			Number of Telephone Calls		
	Total	Business	Residence	Residence Extension	Total	Local	Long Distance
Year	000s	000s	000s	000s	Millions	Millions	Millions
1946	2,026	647	1,379	62	3,559.00	3,484.20	74.8
1947	2,231	712	1,519	81	3,843.30	3,760.60	82.7
1948	2,452	778	1,674	95	4,117.20	4,025.30	91.9
1949	2,700	846	1,854	115	4,559.30	4,454.00	105.3
1950	2,917	902	2,015	128	5,012.60	4,894.70	117.9
1951	3,114	957	2,157	141	5,273.60	5,146.20	127.4
1952	3,352	1,017	2,335	155	5,609.70	5,483.00	126.7
1953	3,606	1,085	2,521	170	6,084.70	5,952.80	131.9
1954	3,860	1,154	2,706	188	6,347.50	6,209.80	137.7
1955	4,152	1,237	2,915	224	6,961.50	6,808.40	153.1
1956	4,499	1,334	3,165	266	7,764.80	7,593.50	171.3
1957	4,827	1,409	3,418	308	8,255.70	8,077.10	178.6
1958	5,118	1,486	3,632	348	8,707.60	8,513.40	194.2
1959	5,439	1,569	3,870	409	9,250.20	9,044.80	205.4
1960	5,728	1,674	4,054	462	9,579.90	9,364.60	215.3
1961	6,014	1,730	4,284	521	10,468.90	10,242.60	226.3
1962	6,329	1,817	4,512	580	10,808.30	10,558.10	250.2
1963	6,657	1,910	4,746	642	11,299.80	11,039.60	260.2
1964	7,019	2,016	5,003	711	11,939.30	11,658.10	281.2
1965	7,445	2,142	5,303	794	12,439.80	12,138.20	301.6
1966	7,893	2,290	5,603	875	13,169.50	12,846.20	323.3
1967	8,358	2,423	5,935	967	13,410.50	13,053.10	357.4
1968	8,818	2,557	6,261	1,084	14,381.60	13,993.60	388
1969	9,296	2,719	6,577	1,188	15,030.90	14,596.60	434.3
1970	9,750	2,854	6,896	1,293	15,895.20	15,436.80	458.4
1971	10,269	2,996	7,273	1,423	16,934.80	16,439.40	495.4
1972	10,987	3,183	7,804	1,667	18,348.90	17,777.00	571.9
1973	11,677	3,428	8,249	1,812	19,054.90	18,396.60	658.3
1974	12,454	3,691	8,763	2,011	20,701.00	19,936.80	764.2
1975	13,165	3,928	9,237	2,194	21,194.10	20,340.60	853.5

Source: *Historical Statistics of Canada*. Statistics cover completed telephone calls only.
Estimates of number of local calls are provided by the larger telephone systems; for each
province and territory, Statistics Canada makes estimates for the remaining systems, on the
assumption that telephones operated by these systems are used to the same extent as those of
the reporting systems.

Table 4
Comparisons Illustrating Bell Growth, 1938 and 1949

Item	1938	1949	Increase
Total telephones	765,855	1,595,184	108%
Calls per day	5,560,000	10,149,000	83%
Annual toll messages	20,493,309	75,798,232	270%
Total telephone plant	$200,360,478	$463,248,589	131%

Source: Compiled from National Transportation Agency Archives, BTC *Exhibits*, box 1, case 955.170, ex. 6.

Table 5
Bell Company Telephones, 1939–1952

Year	Total Telephones	Increase
1939	785,455	–
1940	824,697	5.0
1941	888,348	7.7
1942	925,414	4.2
1943	956,113	3.3
1944	987,530	3.3
1945	1,050,113	6.3
1946	1,177,924	12.2
1947	1,306,975	11.0
1948	1,455,956	11.4
1949	1,595,184	9.6
1950	1,718,200	7.7
1951	1,818,200	5.8
1952	1,908,200	4.9

Total Percentage Increases between 1939 and 1952: 142.9

Source: See table 4, ex. 7.

Table 6
Bell Construction Program, 1939–1952

Year	Gross Construction	Plant Retired	Net Additions
1939	14,850,500	9,982,649	4,867,851
1940	17,418,186	11,823,541	5,594,645
1941	19,523,360	9,862,530	9,660,830
1942	16,750,490	7,581,596	9,168,894
1943	9,511,926	5,182,912	4,329,014
1944	11,551,831	5,106,661	6,445,170
1945	17,990,066	6,185,723	11,804,343
1946	33,798,373	8,057,254	25,741,119
1947	58,659,575	10,371,360	48,288,215
1948	82,575,944	13,346,985	69,228,959
1949	83,848,019	17,661,189	66,186,830
1950	83,214,000	19,578,000	63,636,000
1951	85,247,000	22,374,000	62,873,000
1952	83,319,000	24,123,000	59,196,000

Percentage Increase in Gross Construction: 461.1%

Source: See table 4, ex. 9.

Table 7
Telephone and Utilities Expenditure Detail by Regional Areas, 1947–1948

	Average Dollar Expenditures per Spending Unit				
Utility	Maritimes $	Quebec $	Ontario $	Prairies $	B.C. $
Gas	1	11	14	12	7
Electricity	22	23	27	29	29
Telephone	11	18	22	14	16

Source: Dominion Bureau of Statistics, *City Family Expenditure,* Cat. No. 62-517 Occasional
(Ottawa, 1961).

Table 8
Telephone and Utilities Expenditure Detail by Regional Areas, 1959

	Average Dollar Expenditures per Spending Unit				
Utility	Maritimes $	Quebec $	Ontario $	Prairies $	B.C. $
Gas	5.10	19.20	33.80	51.20	38.20
Electricity	87.50	61.80	69.20	63.10	111.00
Telephone	53.60	64.60	60.70	51.40	61.80

Source: See table 7.

Table 9
Percentage Increase in Utilities Expenditure, 1947 and 1959

Utility	Maritimes $	Quebec $	Ontario $	Prairies $	B.C. $
Gas	410	75	141	327	446
Electricity	298	169	156	118	283
Telephone	387	259	176	267	286

Source: See table 7.

Table 10
Telephones in Operation by Type of Switchboard, 1939–1960

Year	Automatic	% Total	Manual	% Total
1939	690,201	49	707,071	51
1945	1,037,015	56	811,779	44
1953	2,486,451	69	1,119,956	31
1957	3,854,630	80	972,505	20
1960	5,018,308	88	709,859	12

Source: DBS, *Telephone Statistics 1960*, Cat. No. 56-210.

Table 11
Telephone Conversations by Type of Call, Canada: 1939–1960

Year	Local (Millions)	Increase %	Long Distance Calls (millions)	Increase %	Per Capita	Increase %
1939	2,743	n/a	31.6	n/a	245	n/a
1945	3,145	15	64.8	105	265	8
1949	4,454	42	105.2	62	336	27
1953	5,953	34	131.9	25	412	23
1957	8,077	36	178.6	35	498	21
1960	9,365	16	215.3	21	537	8

Source: See table 10.

Table 12
Growth in IBM Sales and Worldwide Service, 1956 and 1965

Type	Units	1956	1965	% Increase
Total Employees	K	72.5	172.45	138
Total Revenue	$B	0.892	3.573	301
Value, Computers in Use	$B	0.244	8.02	3187

Source: Montgomery Phister, *Data Processing Technology and Economics*, rev. ed. (Santa Monica: Santa Monica Publishing Company, 1976), p. 224.

Table 13
Total Worldwise Assets, IBM, 1955 to 1973

Year	Value ($B)	Year	Value ($B)
1956	0.951	1965	3.745
1957	n/a	1966	4.661
1958	n/a	1967	5.594
1959	n/a	1968	6.743
1960	1.976	1969	7.39
1961	2.301	1970	8.539
1962	n/a	1971	9.576
1963	n/a	1972	10.972
1964	3.309	1973	12.289

Source: See table 12, p. 197.

Table 14
Products Shipped by Office Machinery Manufacturers in Canada, 1961–1970
(by value, in thousands of dollars)

Year	Adding Machines	Office Machines	Typewriters	Total	% Change
1961	1,509	45,439	11,828	58,776	–
1962	1,361	61,808	18,173	81,342	38.4
1963	4,849	63,287	17,589	85,725	5.4
1964	1,075	69,174	20,391	90,640	5.7
1965	1,447	61,049	24,519	87,015	−4.0
1966	2,629	99,908	27,033	129,570	48.9
1967	1,156	104,069	20,986	126,211	−2.6
1968	1,163	88,484	22,537	112,184	−11.1
1969	n/a	n/a	n/a	n/a	
1970	10,688	113,308	33,957	157,953	40.8

Total Product Shipments (Value): 931,416
Average Growth in Shipments, Selected Goods: 13.7%

Sources: Compiled from Dominion Bureau of Statistics, *Products Shipped by Canadian Manufacturers*, Manufacturing and Primary Industries Division, Cat. No. 31-211, various years.
Notes: This table demonstrates the anemic performance of the Office Machinery Manufacturing Industry. Computers were not listed as a separate category until 1971, when a reclassification was implemented. Any computers manufactured would likely have been subsumed under that "Office Machines" rubric. The average growth of office machinery manufactures in the period was only 12.6%.

Table 15
Canadian Computer Equipment, Production and Trade, 1964–1972
(in millions of Canadian dollars)

Year	Output	Imports	Exports	Balance	% Change
1964	45	45	32	−13	–
1966	60	105	33	−72	453.8
1968	65	121	41	−80	11.1
1970	110	218	88	−130	62.5
1972	170	321	160	−161	23.8

Source: Statistics Canada, *Trade of Canada,* various years.

Table 16
Office and Store Machinery Manufacturers: Canada, 1957–1975

Year	Establishments	Employees	Factory Shipments ($'000s)	Shipments % Change	Value Added $'000s	Value Added % Change
1957	21	4,035	65,561	–	–	–
1958	24	3,990	57,461	−12.4	–	–
1959	23	3,270	59,015	2.7	–	–
1960	22	3,359	73,767	25.0	37,270	
1961	24	7,759	88,718	20.3	105,775	183.8
1962	24	8,221	96,525	8.8	126,532	19.6
1963	20	8,906	97,903	1.4	129,514	2.4
1964	24	9,044	109,848	12.2	149,833	15.7
1965	23	9,899	105,096	−4.3	166,404	11.1
1966	27	10,171	148,788	41.6	200,854	20.7
1967	29	12,115	163,740	10.0	229,084	14.1
1968	29	12,733	139,006	−15.1	258,180	12.7
1969	27	14,005	154,997	11.5	297,640	15.3
1970	31	15,769	190,571	23.0	361,430	21.4
1971	31	8,696	202,448	6.2	122,327	−66.2
1972	32	10,134	261,822	29.3	152,344	24.5
1973	38	10,866	286,892	9.6	183,090	20.2
1974	43	11,397	246,226	−14.2	240,544	31.4
1975	39	9,613	372,575	51.3	245,131	1.9

Sources: Compiled from DBS, *Office and Store Machinery Manufactures,* Annual Census of Manufactures, DBS, Cat. No. 43-216, 1960–1980.
Notes: SIC 318 includes accounting, adding, addressing, and mailing machines, cash registers, duplicating machines, scales and balances, typewriters, inter alia. Computer-related manufacturers cannot be segregated from the data, especially since certain of these goods come under "Other Office and Business Machinery" to avoid disclosure of individual firms. Establishments counted here include the manufacturing. Output of IBM Canada, among others. Value added is net output as measured by gross output less purchased commodity inputs used. Value of shipments cited only include value of shipments of goods of own manufacture.

Table 17
Bell Canada – Increases in Total Telephones, 1959–1975

Year	In Service	Increase	% Increase
1959	3,330,877	–	
1960	3,515,007	184,130	6
1961	3,695,107	180,100	5.1
1962	3,890,630	195,523	5.3
1963	4,090,102	199,472	5.1
1964	4,312,577	222,475	5.4
1965	4,577,573	264,996	6.1
1966	4,868,392	290,819	6.4
1967	5,152,101	283,709	5.8
1968	5,450,782	298,681	5.8
1969	5,752,820	302,038	5.5
1970	6,007,507	254,687	4.4
1971	6,295,258	287,751	4.8
1972	6,742,184	446,926	7.1
1973	7,102,308	360,124	5.3
1974	7,518,505	416,197	5.9
1975	7,907,439	388,934	5.2

Source: CRTCA, CTC Exhibits, Bell Canada 1975, Rate Application, Ex. B-75-25.
Notes: 1975 estimated.

Table 18
Canadian Personal Expenditures on Consumer Goods and Services,
1960–1977

Year	Personal Expenditures Goods and Services (1968 dollars)	% Change
1960	101,455	–
1961	102,669	1.2
1962	108,009	5.2
1963	112,802	4.4
1964	119,203	5.7
1965	126,425	6.1
1966	133,092	5.3
1967	138,425	4.0
1968	144,642	4.5
1969	152,075	5.1
1970	155,116	2.0
1971	164,327	5.9
1972	176,372	7.3
1973	189,897	7.7
1974	200,889	5.8
1975	210,369	4.7
1976	224,105	6.5
1977	231,190	3.2

Source: Statistics Canada, *Canadian Economic Observer, Historical Supplement,* 1992.

Table 19
Summary of Selected Telephone Statistics: Canada, 1966–1975

	Telephone Calls				
Year	Local '000	Toll '000	Total '000	Per Telephone	Per Capita
1966	12,846,175	323,325	13,169,500	1,668	650
1967	13,053,115	357,414	13,410,529	1,605	665
1968	13,993,601	388,007	14,381,608	1,631	687
1969	14,596,659	434,392	15,031,051	1,617	707
1970	15,436,847	458,397	15,895,244	1,630	737
1971	16,439,365	495,454	16,934,819	1,649	779
1972	17,776,963	571,944	18,348,907	1,670	835
1973	18,396,642	658,248	19,054,890	1,631	854
1974	19,936,758	764,428	20,701,186	1,662	914
1975	20,340,605	853,504	21,194,109	1,610	922

Source: Statistics Canada, Telephone Statistics, Cat No. 56-203, 1975.

Table 20
Telephones by Type of Service and per 100 Population, Selected Provinces, 1975

Province	Individual Lines		Two-Party Lines		Four-Party Lines		4+ Party Lines		PBX Telephones		Extension Telephones		Total Telephones	Per 100 Population
	Res.	Bus.	Res.	Bus.	Res.	Bus.	Res.	Bus.	Res.	Bus.	Res.	Bus.		
Quebec	1,592,710	274,294	142,682	3,077	16,422	741	127,803	4,043	46	350,751	509,798	241,330	3,423,405	55
Ontario	2,207,281	404,290	243,435	3,012	7,781	313	173,016	5,757	54	532,432	932,104	327,437	5,091,709	61
Saskatchewan	211,748	49,061	6	2	–	–	67,246	2,515	–	42,083	74,437	30,972	476,207	51
British Columbia	563,176	132,254	156,219	30	3,522	2	77,992	3,003	–	165,663	246,914	69,798	1,485,228	60
Canada	5,769,742	1,086,725	590,189	8,384	122,501	5,546	559,893	20,146	100	1,378,372	2,194,210	904,922	13,165,010	57

Source: See table 19.

Table 21
Bell Canada Selected Revenue Statistics, 1967 to 1975

Year	Total Operating Revenues	% Change	Local Service Revenues '000	% Change	Long Distance Revenues '000	% Change
1967	671,830	n/a	409,989		258,944	
1968	725,945	8.1	437,554	6.7	284,713	10.0
1969	803,430	10.7	471,127	7.7	327,809	15.1
1970	867,374	8.0	504,363	7.1	359,480	9.7
1971	924,901	6.6	536,853	6.4	381,689	6.2
1972	1,015,337	9.8	577,901	7.6	428,605	12.3
1973	1,127,724	11.1	622,082	7.6	495,509	15.6
1974	1,252,137	11.0	675,574	8.6	565,599	14.1
1975	1,406,181	12.3	732,551	8.4	622,398	10.0

Source: CRTCA, CTC Exhibits, Bell Canada, 1975 (B-75-327, 328, 335).

Table 22
Telephone Statistics, Selected Countries, 1966 and 1975

Country	1966	1975	Change	Per 100 Population
Canada	7,445,071	12,454,331	67.3	57.15
France	6,116,700	12,405,000	102.8	26.2
Fed. Rep. Germany	8,802,166	18,767,033	113.2	31.7
Italy	5,980,702	13,695,006	129.0	25.88
Japan	13,998,831	41,904,960	199.3	40.47
United Kingdom	10,704,000	20,342,457	90.0	37.51
United States	93,656,000	143,972,000	53.7	69.49

Source: AT&T, The World's Telephones as of January 1, 1976.

Table 23
Comparison of US and Canadian Government Procurement of Computers, 1970

Company	Number of Machines (%)		Value (%)	
	USA	Canada	USA	Canada
IBM	26.0	35.9	40.3	55.0
CDC	8.4	2.2	17.3	1.9
UNIVAC	15.6	7.6	11.8	7.7
Honeywell	8.5	15.1	7.4	5.0
RCA	3.1	1.1	7.0	1.6
Burroughs	4.7	12.0	5.4	13.3
XDS	5.2	8.7	4.0	7.6
DEC	13.0	7.6	1.6	3.8
Others	16.5	9.7	4.6	4.1

Source: NA, Records of the Treasury Board, RG 55, Acc. 1986-87/379, box 32, file 7148-0, pt. 1, "Relationship of Federal Government Procurement of EDP Equipment and Services to Overall National Policies," Memorandum to Cabinet, 21 February 1972.

Table 24
Annual Summary of us Business and High Technology Imports to Canada, 1968–1980

Year	Electronic Computers '000	Growth %	Office Machines '000	Growth %	Semiconductors and Tubes '000	Growth %	Total All Imports '000	Growth %
1968	98,497		156,359		38,283		9,050,565	
1969	151,154	53.5	223,397	42.9	50,472	31.8	10,243,242	13.2
1970	163,875	8.4	259,881	16.3	49,677	-1.6	9,917,044	-3.2
1971	160,501	-2.1	297,603	14.5	51,246	3.2	10,950,628	10.4
1972	196,227	22.3	371,620	24.9	78,087	52.4	12,878,180	17.6
1973	261,984	33.5	398,198	7.2	104,819	34.2	16,502,016	28.1
1974	317,117	21.0	475,972	19.5	107,457	2.5	21,386,714	29.6
1975	324,519	2.3	520,241	9.3	61,274	-43.0	23,641,313	10.5
1976	408,500	25.9	610,802	17.4	78,146	27.5	25,801,282	9.1
1977	531,071	30.0	680,921	11.5	91,102	16.6	29,841,436	15.7
1978	779,119	46.7	901,238	32.4	140,383	54.1	35,432,780	18.7
1979	1,015,190	30.3	1,155,770	28.2	343,013	144.3	45,571,223	28.6
1980	1,559,590	53.6	1,711,847	48.1	517,069	50.7	48,613,643	6.7
Totals	5,967,344		7,763,849		1,711,028		299,830,066	
Avg. Growth		27.1		22.7		31.1		15.4

Source: "Merchandise Imports from U.S.A., by commodities, based on the Standard Commodity Classification Monthly, in Thousands of Dollars," *Trade of Canada Imports* Cat. No. 65-007 (Ottawa: Statistics Canada), various years. See also "Merchandise Imports from USA, Country of Origin, by Commodities, Based on the Harmonized System Classification, Monthly, in Thousands of Dollars," ibid., Catalogue No. 65-006.

Table 25
Computer Installations in Canada, 1968 to 1974, Various Provinces

Year	Ontario	Quebec	Saskatchewan
1968	81	410	41
1970	136	603	57
1972	2,279	939	91
1974	2,400	1,035	90

Source: Canadian Information Processing Society, Canadian Computer Census, 1974 (Toronto, 1975), 15. These figures do not include computers rented for less than $1,000 per month.

Table 26
Imports of Electronic Computers and Parts into Canada, 1960–1980

Year	Value '000	Annual Growth %
1960	3,149	
1961	8,505	170.1
1962	13,359	57.1
1963	16,359	22.5
1964	30,311	85.3
1965	50,511	66.6
1966	93,495	85.1
1967	115,902	24.0
1968	108,606	−6.3
1969	160,527	47.8
1970	176,290	9.8
1971	184,161	4.5
1972	212,418	15.3
1973	276,813	30.3
1974	333,116	20.3
1975	343,350	3.1
1976	428,323	24.7
1977	562,074	31.2
1978	852,564	51.7
1979	1,103,454	29.4
1980	1,652,774	49.8
Total Value of Computer Imports, 1960 to 1980	6,726,060	
Average Annual Growth in Imports		41.1

Source: Compiled from data in Trade of Canada, various years.

Table 27
Total Operating Revenue, Canadian Computer Service Industry, 1972–1979 ($'000s)

Year	Under $100K Value	Companies with a Revenue of		$2M+ Value	Total Value	Change % Total
		$100–$499 Value	$500K–$1.9M Value			
1972	6,171	21,150	54,885	425,789	507,994	
1973	6,124	26,731	42,645	514,212	589,712	16.1
1974	n/a	n/a	n/a	n/a	n/a	
1975	6,553	40,012	61,428	177,688	285,681	−51.6
1976	9,102	42,148	73,374	201,907	326,531	14.3
1977	10,696	51,128	83,760	270,464	416,048	27.4
1978	12,248	60,262	103,456	355,820	531,786	27.8
1979	10,218	66,987	125,273	435,560	638,038	20.0

Source: Statistics Canada, Merchandising and Services Division, *The Computer Service Industry,* Catalogue Number 63-222, 1972–1979.

Table 28
Bell Assets, Liabilities, Operating Revenues and Expenses, 1945–1975
(in millions of dollars)

Year	Assets	% Change	Liabilities	% Change	Operationg Revenue	% Change	Operating Expenses	% Change
1945	294	–	294	–	69	–	43	–
1946	326	11.0	326	11.0	77	11.3	54	24.7
1947	379	16.2	379	16.2	85	10.5	63	16.6
1948	447	18.0	447	18.0	96	12.2	73	15.9
1949	515	15.1	515	15.1	108	12.5	88	20.3
1950	577	12.0	577	12.0	128	19.0	102	15.3
1951	639	10.8	639	10.8	159	24.0	115	13.4
1952	716	12.1	716	12.1	184	15.9	127	9.8
1953	843	17.6	843	17.6	202	9.5	139	10.0
1954	914	8.4	914	8.4	219	8.6	154	10.5
1955	1,030	12.8	1,030	12.8	245	11.6	173	12.7
1956	1,183	14.8	1,183	14.8	274	11.9	196	13.3
1957	1,355	14.5	1,355	14.5	303	10.6	220	12.3
1958	1,493	10.2	1,493	10.2	329	8.6	239	8.6
1959	1,674	12.1	1,674	12.1	377	14.5	256	7.1
1960	1,815	8.4	1,815	8.4	405	7.5	270	5.5
1961	1,983	9.3	1,983	9.3	434	7.1	282	4.5
1962	2,197	10.8	2,197	10.8	471	8.6	302	6.9
1963	2,347	6.8	2,347	6.8	503	6.8	326	7.9
1964	2,552	8.7	2,552	8.7	543	7.9	343	5.4
1965	2,719	6.5	2,719	6.5	593	9.2	372	8.4
1966	2,460	–9.5	2,460	–9.5	645	8.8	407	9.2
1967	2,652	7.8	2,652	7.8	702	8.8	430	5.8
1968	2,863	8.0	2,863	8.0	758	8.0	464	7.8
1969	3,108	8.6	3,108	8.6	842	11.0	530	14.3
1970	3,370	8.4	3,370	8.4	937	11.2	578	9.1
1971	3,726	10.6	3,726	10.6	1,019	8.8	640	10.6
1972	4,581	23.0	4,581	23.0	1,836	80.2	1,373	114.7
1973	5,139	12.2	5,139	12.2	2,102	14.5	1,574	14.6
1974	5,281	2.8	5,281	2.8	1,693	–19.4	1,192	–24.3
1975	6,588	24.8	6,588	24.8	1,967	16.2	1,390	16.6
Average Annual Growth		11.1		11.1		12.5		13.6

Source: Bell Annual Reports, 1945–1975

Table 29
Bell Net Income and Dividends, 1945–1975

Year	Net Income '000	Change %	Total Dividends '000	Change %	Dividends per Share
1945	8,304	–	6,898	–	$2.00
1946	8,274	–0.4	7,299	5.8	$2.00
1947	9,702	17.2	8,822	20.9	$2.00
1948	10,577	9.0	10,187	15.5	$2.00
1949	7,841	–25.9	12,485	22.6	$2.00
1950	10,324	31.7	13,535	8.4	$2.00
1951	16,203	57.0	15,614	15.4	$2.00
1952	22,570	39.3	18,272	17.0	$2.00
1953	26,849	19.0	20,264	10.9	$2.00
1954	28,549	6.3	23,378	15.4	$2.00
1955	31,978	12.0	26,309	12.5	$2.00
1956	34,949	9.3	29,147	10.8	$2.00
1957	36,037	3.1	33,481	14.9	$2.00
1958	38,899	7.9	36,180	8.1	$2.00
1959	50,284	29.3	41,575	14.9	$2.00
1960	53,512	6.4	46,652	12.2	$2.20
1961	57,691	7.8	50,786	8.9	$2.20
1962	62,285	8.0	53,968	6.3	$2.20
1963	68,294	9.6	58.189	7.8	$2.20
1964	77,739	13.8	63,005	8.3	$2.20
1965	86,525	11.3	65,129	3.4	$2.20
1966	91,963	6.3	75,859	16.5	$2.43
1967	108,537	18.0	85,941	13.3	$2.50
1968	114,329	5.3	87,358	1.6	$2.50
1969	113,696	–0.6	88,949	1.8	$2.50
1970	133,262	17.2	99,155	11.5	$3.51
1971	147,290	10.5	110,021	11.0	$3.72
1972	175,486	19.1	112,049	1.8	$4.30
1973	205,371	17.0	120,500	7.5	$4.99
1974	224,436	9.3	133,354	10.7	$3.12
1975	317,362	41.4	160,263	20.2	$3.44
Average Annual Growth		13.9		11.2	

Source: Bell Annual Reports, 1945–1975

Table 30
International Comparison of Telecom Traffic: Per Capita Calls in Selected Countries,
1971–1980

	Canada	USA	UK	Germany	Italy	Japan
1971	780.1	809.3	–	189.4	194.2	–
1972	840.4	846.8	218.3	211.6	206.9	414.9
1973	857.0	883.7	236.6	216.5	158.6	415.9
1974	917.2	930.6	256.0	222.2	177.8	414.4
1975	926.1	951.5	271.0	225.5	177.2	411.2
1976	959.5	961.4	285.5	246.5	192.9	408.9
1977	993.6	999.4	309.7	262.4	200.3	357.2
1978	1020.3	1050.8	335.3	286.1	211.1	366.4
1979	1054.1	1082.9	355.6	310.6	232.2	–
1980	1114.2	1122.2	361.0	352.1	242.4	–

Source: Compiled from tables in ITU, Yearbook of Common Carrier Telecommunication Statistics, 9th ed.
(Chronological Series 1971–1980)

Table 31
Per Capita Income from All Telecommunications Services, Selected Countries
1971–1980 (current US dollars)

Year	Canada	USA	UK	Germany	Italy	Japan
1971	83.80	98.52	–	56.23	25.68	–
1972	93.21	110.20	46.05	73.98	29.85	39.27
1973	104.67	123.07	48.80	90.88	33.72	52.44
1974	119.20	135.42	58.43	117.44	36.06	58.31
1975	91.18	148.18	93.83	122.01	44.97	59.64
1976	155.86	166.58	91.27	148.54	42.89	64.83
1977	162.42	185.48	89.45	182.87	55.23	78.40
1978	172.32	206.56	108.17	229.90	63.88	137.96
1979	200.38	225.05	131.69	252.38	74.76	155.01
1980	215.66	248.49	168.19	215.72	82.61	164.09

Source: See table 30.
Note: Income consists of all telecommunication revenue earned during the financial year under
review. This may include income from subscribers, other national and foreign telecommunication administrators, governments, etc. after deduction of the share of this income to be paid
to other administrations for outgoing telecommunication traffic.

Notes

INTRODUCTION

1 The literature on regulation is considerable. The more outstanding
  contributions on utilities regulation are: Kahn, *The Economics of Regulation*, 2 vols.; Stigler, *The Citizen and the State: Essays on Regulation*;
  Schmalensee, *The Control of Natural Monopolies*; Wilson, ed., *The Politics
  of Regulation*; Anderson, *Regulatory Politics and Electric Utilities*; McGraw,
  ed., *Regulation in Perspective*; McGraw, *Prophets of Regulation*; Bickers,
  "The Politics of Regulatory Design; Telecommunications Regulation in
  Historical and Theoretical Perspective"; Braeutigam and Owen, *The Regulation Game*; Braeutigam, *The Regulation of Multiproduct Firms: Decisions
  on Entry and Rate Structure*; and Skowronek, *Building a New American
  State*. More dated but nonetheless useful contributions are: Collins, *A
  Voice from Afar*, and Ogle, *Long Distance Please*. A notable contribution in
  political science has been Frederick and Martha Fletcher's "Communi-
  cations and Confederation: Jurisdiction and Beyond," in Byers and
  Reford, eds., *Canada Challenged* and especially the several contributions
  of Richard Schultz, including *Federalism and the Regulatory Process*
  (Montreal, 1979), and *Federalism, Bureaucracy, and Public Policy*. See also
  Janisch, *The Regulatory Process of the Canadian Transport Commission*;
  "The North American Telecommunications Industry from Monopoly to
  Competition," in Button and Swann, eds., *The Age of Regulatory Reform*.
  The studies generated by the Telecommission, a Department of Com-
  munications task force in 1969–71 are also helpful. See also Stanbury,

ed., *Telecommunications Policy and Regulation*. For a technical history of the Canadian digital network, see Dewalt, *Building a Digital Network*.

2 The literature on the computer industry in general focuses, not surprisingly, on the United States. See Beniger, *The Control Revolution*; Cortada, *Before the Computer*; Fisher, McKie, and Mancke, IBM *and the U.S. Data Processing Industry*; Flamm, *Creating the Computer*; in the British context, see Hendry, *Innovating for Failure*. There is no major study of the Canadian computer industry.

On technological history, see Hughes, "The Order of the Technological World," in Hall and Smith, eds., *History of Technology, vol. 5*; *Networks of Power*; and "The Dynamics of Technological Changes: Salients, Critical Problems and Industrial Revolution"; Dosi, Giannetti, and Toninelli, eds., *Technology and Enterprise in a Historical Perspective*. See also David Landes's *Unbound Prometheus*.

3 In *re Regulation and Control of Radio Communication in Canada*, Judicial Committee of the Privy Council, [1932], A.C. 304.

CHAPTER ONE

1 Nelles, and Armstrong, *Monopoly's Moment: The Organization and Regulation of Canadian Utilities 1830–1930*, 188–91.

2 Federal power is derived from s. 92(10)(a) of the Constitution Act, 1867, and Judicial Committee of the Privy Council (hereafter JCPC) in *Toronto Corporation* v. *Bell Telephone Company of Canada* [1905], A.C. 52. Cf. Buchan and Johnston, "Telecommunications Regulation and the Constitution: A Lawyer's Perspective," in Buchan et al., *Telecommunications Regulation and the Constitution*, 115–66.

3 Hughes, "The Order of the Technological World," in Hall and Smith, eds., *History of Technology*, vol. 5, 2; Hughes, *Networks of Power*, 363.

4 See Lears, "A Matter of Taste: Corporate Cultural Hegemony in a Mass-Consumption Society," in May, ed., *Recasting America*.

5 See Latour, *Science in Action*; also Michael L. Smith, "Representations of Technology at the 1964 World's Fair," in Fox and Lears, eds., *The Power of Culture*, 226, and Melling and Barry, "The Problem of Culture: An Introduction," in Melling and Barry, eds., *Culture in History*.

6 67 CRTC 1, p. 7; Bell Canada Archives (hereafter BCA) Information Bulletins (hereafter IB), vol 2. 1949–52, no. 71, 20 November 1950; 71 CRTC 286 (1954). Re Report No. 161, Windsor, Ont., Exchange; and Re Report No. 150, Ottawa, Exchange. See also Order no. 84101 … file no. 46638.1, 9 July 1954.

7 Kerr, *The Board of Transport Commissioners for Canada. A Review of Its Constitution, Jurisdiction and Practice* (Ottawa, The Board, 1959), 26. Cabinet power derived from s.53(1) of the Railway Act and order-in-

council (hereafter OIC) of 15 October 1918. See ss. 56(2) and 56(3) of the Railway Act of 1906; OIC, 17 June 1927 and 25 February 1933.

8 34 Canadian Railway Cases (hereafter CRC) 1, 21 February 1927 see also 16 Judgements, Orders, Regulations and Rulings (hereafter JOR&R) 230, 67 CRTC 1, p. 60, and 34 CRC 1, p. 34.

9 Cf. Re B.C. Tel. Company... 66 CRTC 7; and 67 CRTC 1, p. 63, 71; CRTC 286, 175; also 34 CRC 1; Circ. no. 267, 18 October 1951.

10 MacPherson was a Liberal member of the BC legislature from 1928 to 1933 and BTC commissioner from 1935 to 1958. Armand Sylvestre was a Liberal MP from 1925 to 1945. Sylvestre was deputy chief commissioner for the greater part of the 1950s.

11 NA, John G. Diefenbaker Papers (hereafter Diefenbaker MSS), MG 26 M VI MFM M-7889, file 313.3G1/774, C.B. Devlin to George Hees, Minister of Transport, 23 July 1959; Hees to Diefenbaker, 8 January 1959. See also Louis St Laurent Papers (hereafter St Laurent MSS), MG 26 L, vol. 93, file D-40-B, H.A. Gosselin to St Laurent, 9 May 1951.

12 St Laurent MSS, vol. 93, file D-40-B; Minister of Transport to St Laurent, 10 August 1950; St Laurent to James Gardiner, 18 August 1950; W.J. Patterson to St Laurent, 20 April 1951.

13 Archives Municipales de Montréal (hereafter AMM), dossier 92699, troisieme série, Conseil Rapports et Documents (hereafter 3ᵉ série), Avocat en Chef de la cité à Me. Louis A Lapointe et Monsieur le President, MM les membres du Comité executif, le 13 janvier 1950. "David-Goliath Setting at Telephone Hearing," in Montreal Gazette, 15 May 1965.

14 Donald Creighton, The Forked Road: Canada 1939–1957 (Toronto, 1976), 107–22. The Wartime Prices and Trade Board, 7 April 1942, National Transportation Agency Archives (hereafter NTAA), BTC Transcripts (hereafter BTCT), box 11, vol. 732, file 36730.2, 11, 12, 13 September 1946. See also BTC, JOR&R, vol. XXXVI, Ottawa, 1 January 1947, La Ligne Téléphonique des Cultivateurs ..., file 3839.511.1; BTCT, box 11, vol. 732, 1578. See also Jean-Guy Rens, L'Empire Invisible. Histoire Des Télécommunications Au Canada. Vol. I: De 1846 à 1956 (Sainte-Foy, Québec, 1993), 451–3. BTCT, box 11, vol. 732, 13855, 13808. Oral Judgement of 12 September 1947.

15 CRTC, Bell Telephone Co. v. Cities of Toronto, Montreal, Ottawa et al., BTC, 15 November 1950. The last increase was granted on 21 February 1927 and reported in 34 CRC 1 and also in 16 JOR&R 230. 67 CRTC 1, p. 6. BTCT, box 26, vol. 838, file 955.170, 3065.

16 AMM, dossier 92699, 3ᵉ série, N.A. Munnoch to Claude Choquette, 8 November 1949; BCA, IB vol 2, no. 17, 19 December 1949. One Toronto woman took advantage of the delay to prepare a submission for the hearing. The board allowed one Sophie Kohen to air her

concerns that her telephone cord was filling her house with gas fumes, and that her phone was being tapped by Nazi vampire telephone operators, who drained the blood of young children when they were not harassing Kohen with questions like "Are you a virgin?" – presumably contrary to Bell's charter. "'That's What It's For,' Reply to Bell's Stand Surplus Being Used Up," in *Globe and Mail*, 9 March 1950; BTCT, box 26, vol. 838, file 955.171, 8 March 1950, 3260–70.

17 "'That's What It's For,' *Globe and Mail*, 9 March 1950. "Les principales villes du Québec et de l'Ontario demandent le rejet de la demande faire par la Cie. Bell," *Le Devoir*, 9 mars, 1950; BCA, IB, 8 March 1950; BTCT, box 26, vol. 838, file 955.170, 3116–7; "La Compagnie bell s'engagerait dans une expansion non immédiatement nécessaire," *Le Devoir*, 13 mars 1950.

18 BCA, IB, 1 June 1950; BTCT, box 27, vol. 846, 5930.

19 BCA, IB, 13 April 1950; BTCT, box 27, vol. 839, 3692 and vol. 858, 9295.

20 BCA, IB, 11 May 1950; ibid., 14 July 1950; "New Bell 'Phone Rates Add $1\frac{1}{2}$ Million a Month," *Financial Post*, 22 June 1950.

21 BTCT, box 29, vol. 855, 9290; vol. 858, 9527.

22 67 CRTC 1, p. 28.

23 City of Toronto Archives (hereafter CTA), Board of Control Minutes (hereafter BCM), 1950, vol. 2, 22 November 1950.

24 BCA, IB, 31 August 1951; AMM, dossier 99024, 3ᵉ série, Claude Choquette à M. L.A. Lapointe et al., 13 septembre 1951.

25 BCA, IB, 26 September 1951; ibid., 26 September 1951; Fernand Jolicœur, "Service public et propriété privée," *Le Travail*, September 1951; "Bell Telephone and Public Ownership," *Financial Times*, 21 September 1951;"For First Time Majority Opposes Gov't Ownership," by Canadian Institute of Public Opinion, reprinted in *Montreal Star* and *Toronto Daily Star*, 24 November 1951. Ten per cent had "No opinion"; BCA, IB, 16 November 1951; AMM, Dossier 99024, 3ᵉ série, Chief City Attorney to Executive Committee, 21 January 1952; and 68 CRTC 127, 13 November 1951.

26 68 CRTC 359, 65;47 JOR&R, 1 April 1952, "In the matter of the Application of The Bell Telephone Company of Canada," dated 31 August 1951, Case 955.171.

27 See for example 67 CRTC 1, 68; Case No. 955.170, 18 July, 28 August, 7 November and 5 December 1951; and file C 955.170, 12 March 1954, 71 CRTC 319 and 320 (1954).

28 BCA, IB, June 1, 1950.

29 46 JOR&R 14, 15 October 1956; BTCT, box 1, vol. 858, 9493. See also *Montreal Gazette* "City Seeks to Check Bell Plea to Increase Telephone Rates" 9 March 1950; 67 CRTC 1, 79. Many company-municipal disagreements came before the board. See 44 JOR&R, 1 November 1954.

BTCT, box 58, vol. 1036, file 44484.24, "In the Matter of the Application of the Bell Telephone Company ..." 28 and 29 January 1959. See RE B.C. Telephone Co. (1950) 66 CRTC 7. On distribution of the rate burden see 34 CRC 1 (1927); 67 CRTC 1 (1950); 71 CRTC 286 (1954); 46 JOR&R 14, 15 October 1956, 244; 71 CRTC 286 (1954).

30 48 JOR&R 13, 1 October 1958, file c.955.170.4, p. 265.

31 See, for example, NA, RG 46 C-II-1, vol. 1670, pts. 6 and 7; BTC, file 46638.1, pt. 6, 1 May 1957; Rens, *L'Empire Invisible*, 366.

32 A Submission to the Royal Commission on Canada's Economic Prospects by Mayor Nathan Phillips, January 1956; J.M. Smith, *Canadian Economic Growth and Development From 1939 to 1955* (Ottawa: The Commission, 1957), 7,44.

33 Transcript of hearings of the RCCEP, testimony of T.W. Eadie, 6 March 1956, 8569, 8572.

34 Testimony of John Kenneth Galbraith before the RCCEP, Toronto, Ontario, 24 January 1956; BCA, *Bell Canada News*, "At the Gate of '58," 25 December 1957.

35 Will, *Canadian Fiscal Policy 1945–1963*, 55, 62; Commons Standing Committee on Railways, Canals & Telegraph Lines, *Evidence*, 28 November 1957; 47 JOR&R 24, 15 March 1958, c. 955.172; "Bell Links Rate Boost to National Defense," *Globe and Mail*, 21 November 1957. Employee expense per telephone rose from $39.50 in 1952 to $42.40 in 1956; "Wages, Costs Justify Higher Rates: Bell" *Globe and Mail*, 20 November 1957.

36 CTA, BOCM, 1957, vol. 2; see BOC Report 23, in Appendix A, City of Toronto, Council Minutes (hereafter TCM), 2225.

37 TCM, BOC Report 28, Appendix A, 2651, 2562. See also AMM, 3ᵉ série, City Attorney to Lucien Hétu Re: The Bell Telephone Company of Canada ... 6 September 1957. "Bell Increase Unjustified, Brief Says," *Globe and Mail*, 19 November 1957; BTCT, box 53, vol. 1007, 11 December 1957, 7495; ibid., vol. 1005, file 955.172, 21 November 1957, 7186, 7186A, 449.

38 TCM, 1957, vol. 2, see BOC Report, 32, Appendix A, 29 November 1957.

39 "Hausse de trois pour cent des taux de téléphone dans l'Ontario et le Québec," *Le Devoir*, 11 janvier 1958.

40 The diminishing balance method carried a greater annual charge in the earlier years of service life and less in the later, resulting in lower income tax payments in the earlier years and higher in the later years. BCA, IB 11, 23 May 1958.

41 See Helliwell, *Taxation and Investment*, 150ff.

42 Ibid.

43 J.G. Diefenbaker Centre, Prince Albert, Saskatchewan, John G. Diefenbaker Papers (hereafter DP), MG 01/XXII, press releases series file:

RE: Increases in Rates by Freight Railroads and Telephone Companies, 26 February 1958. The government suspended the increases twice, since the Diefenbaker cabinet was labouring mightily under a heavy political and legislative agenda.

44 Smith, *Gentle Patriot*, 56; NA, RG 19, vol. 4411, file 9045-03-2, pt. 1, Memorandum … on Behalf of the Bell Telephone Company of Canada, n.d., probably March 1958.

45 DP, XXII/105, 29 April 1958.

46 "The Real Reason," editorial, *Globe and Mail*, 1 May 1958. "Ottawa annule la hausse des tarifs de téléphone" *Le Devoir* 30 avril 1958; Cabinet Refuses Boosts for Phone, Rail Rates," *Globe and Mail*, 30 April 1958.

47 "Phone, Rail Boosts Out," Montreal *Gazette*, 30 April 1958.

48 DP, VI/R/1064, 386605, Memorandum for the Prime Minister 16 May 1958; ibid., 386601, T.W. Eadie to Diefenbaker, 20 June 1958.

49 BCA, IB, 25 June 1958; J.J. Frawley to BTC, in 48 JOR&R 18, 15 December 1958, 395.

50 DP, VI/7753, 386499, Stanley Petrie to Diefenbaker, 24 July 1958; BTCT, box 56, vol. 1027, file 955.173, 15 September 1958, 7524.

51 CTA, Carroll to Mooney, 3 July 1958.

52 DP, VI/7753, 386513, "The Problem of Deferred Tax Credits," Bell Canada pamphlet, Toronto area, June, 1958; ibid., 386601, W.R. Brunt to Diefenbaker, 15 July 1958.

53 *Globe and Mail*, 3 July 1958; *Financial Times*, 25 July 1958.

54 "High Pressured, Aldermen Claim, Declare War on Phone Rate Rise," *Globe and Mail*, 22 July 1958; "When Sauce for the Goose is NOT Sauce for the Gander," *Financial Times*, 25 July 1958.

55 Editorial, *Le Droit*, 23 July 1958.

56 BCA, IB, 14 October 1958.

57 Ibid., 24 October 1958; DP, File: Re: Suspension … vol 2, 24 October 1958; BCA, IB, 6 November 1958. Memorandum Re: Bell Telephone Company 13 October 1958. DP, p. 386476.

58 NA, RG 19, vol. 441, file 9045-03-02, pt. 3, J.A. Hayes to S.S. Reisman Re: Governor-in-Council Hearing on Bell Telephone Rate Increases, 19 November 1958;BCA, IB, 27 November 1958.

59 DP, VI/R/1064, 386600, Gowan T. Guest to Hon. George Hees, 20 November 1959; ibid., 386598-9, Hees to Guest, 27 November 1959.

60 Neville Nankivell, "Power and Public Utilities," *Financial Post*, 6 May 1961 and "Big Spending Program under Way by Utilities,""Bell Phone Net Profit Rises 16.9%"; "Earnings Rise Steadily," *Financial Post*, 5 May 1962;"Magic of Communications In Bell Telephone Exhibit," *Financial Post*, 22 September 1962; W.C. MacPherson, "Telephone's Versatility Grows as Machines Talk to Machines," *Financial Post*, 18 May 1963.

61 54 JOR&R 4, 13 January 1964; 53 JOR&R 4, 31 July 1963; NA, RG 46-C-II-1, vol. 1716, Exhibit IW&C-1, Memorandum on Bell–Northern Relationships."

62 Exhibit no. B-54, "Memorandum on Bell-Northern Relationships," 9–10, 12, 54, 56.

63 NA, RG 46–C-II-1, vol. 1716, R.M. Davidson to D.H.W. Henry re Bell–Northern Electric, 1 February 1965.

64 "Bell Profits Ruling Is Proving Headache for Commissioners," *Globe and Mail*, 4 June 1965; "Rates Must Come Down if Profit Up, Bell Told," ibid., 5 May 1966.

65 Robert Rice, "Bell and Municipalities Prepare for Battle on Proposed New Profit Yardstick," *Globe and Mail*, 4 May 1965; "Bell's Real Plea: Don't Fence Us In," *Financial Post*, 5 December 1964.

66 BTCT, box 74, vol. 1118, Inquiry Re: Bell Telephone Company of Canada, 2884. See also NA, RG 46, files 1113-1118; Amy Booth, "New Wooing of 'Mother Bell'," *Financial Post*, 23 January 1965. Calls for nationalization emanated especially from Quebec. Author's interview with Eric Kierans, 4 June 1992. Cf "Bell Rings Out against Critics Seeking to Nationalize Utility," *Financial Post*, 20 March 1965.

67 BTCT, box 74, vol. 1121, 21 June 1965.

68 "David-Goliath Setting at Telephone Hearing," *Montreal Gazette*, 15 May 1965; "Utilities Ombudsman Urged at Bell Inquiry," *Globe and Mail*, 30 June 1965.

69 "Bell May Have Trouble Finding Investors if not Allowed Profit Raise, Witness Says," *Globe and Mail* 12 May 1965; BTCT, box 74, vol. 1120, 16 June 1965, 3627, 3630.

70 The company earned 6.6 per cent on capital in 1965; "Bell Sees Decision as 'Step Forward,'" *Montreal Star*, 5 May 1966.

CHAPTER TWO

1 See Granatstein, *The Ottawa Men*, 129.

2 On state intervention, see Tupper and Doern, "Public Corporations and Public Policy in Canada," in Tupper and Doern, eds., *Public Corporations and Public Policy*, 10.

3 See Commonwealth Telecommunications Board (hereafter CTB), *1st Annual Report*, 1951, A3. Cf. Canada, House of Commons, *Debates*, Overseas Telecommunication Act, 30 September 1949, 387; Public Record Office (UK), DO 35/1805, File WQ 351/1.

4 DEA/66s, High Commissioner in Great Britain to Secretary of State for External Affairs (SSDEA), 28 January 1942. See also replies in NA, WLMK Papers, vol. 329, SSDEA to High Commissioner in Great Britain,

29 January 1942. Quoted in Hilliker, ed., *Documents on Canadian External*, vol. 9 (hereafter *DCER*), 966ff.

5 DEA/5615–40, N.A. Robertson to W.A. Riddell, High Commissioner in New Zealand, 19 November 1943; John Holmes, Memorandum on Mr. Curtin's Proposals for an Empire Council, n.d.

6 DEA/6231–4, Dominions Secretary to Secretary of State for External Affairs, London, 16 September 1944.

7 Sir Campbell Stuart, Memorandum of 13 February 1945, quoted in Hugh Barty-King, *Girdle Around the Earth. The Story of Cable and Wireless and Its Predecessors to Mark the Group's Jubilee, 1929–1979* (London, 1979), 287.

8 See PRO, secret file, British Embassy, Washington G 285/1, Doc Reference FO 115 3571. Quoted also in Barty-King, *Girdle Around the Earth*, 297.

9 PRO, DO 35/1805, WQ 351/1, Sir Edward Wilshaw, to Lord Cranborne, Secretary of State for Dominion Affairs, 15 October 1943.

10 Ibid., Secret, Commonwealth Communications Council, Second Interim Report to the Governments, London, 23 May 1944, Sir E. Machtig to Secretary of State, 25 April 1944.

11 *DCER*, vol. 10, 1271.

12 Ibid., "Memorandum of the Interdepartmental Committee ...," 1271.

13 DEA/6327–40, Memorandum from Secretary, ICTP to CWC, Ottawa, 26 February 1945; ibid., USSDEA to High Commissioner of Great Britain, 2 March 1945.

14 Ibid., Prime Minister to Head, British Delegation on Telecommunications, Secret, Ottawa, 16 March 1945.

15 Ibid., Memorandum from Secretary, ICTP, "Commonwealth Telecommunications: Canadian Policy," March 1945.

16 PCO, vol. 24, "Memorandum from Secretary, Cabinet Committee on Reconstruction to Minister of Reconstruction," Ottawa, 6 July 1945.

17 DEA/6231–40, acting chairman, ICTP, to Minister of Reconstruction Ottawa, 6 July 1945. Empire/foreign traffic signified traffic originating in the Empire, but having a non-Empire destination. Foreign/Foreign Traffic was traffic that originated and terminated in a foreign country, but was routed through a Commonwealth country.

18 Ibid., C.D. Howe to Under-Secretary of State for External Affairs, Norman Robertson, Ottawa, 14 July 1945.

19 Ibid., Secret Report of Canadian Delegation to Commonwealth Telecommunications Conference in London, 16 July to 3 August 1945, London, 17 August 1945.

20 Ibid., Confidential Memorandum from Secretary, Interdepartmental Committee on Telecommunications Policy, to Cabinet, Ottawa, 24 January 1946, Commonwealth Telecommunications Conference, 1945; ibid.,

7767–40, Massey, USSDEA, to Robertson, High Commissioner in Great Britain, Secret, Ottawa, 4 March 1946.

21 Ibid., Secret Memorandum from Secretary, ICTP to Cabinet, 2 November 1945, "US-Commonwealth Telecommunications Conference; Canadian Policy."

22 Ibid., approved 7 November 1945.

23 Ibid., Chairman, Delegation to Bermuda Telecommunications Conference, to Chairman, ICTP, Secret Despatch, Ottawa, 13 December 1945.

24 Canada, *Treaty Series, 1945*, 28 February 1946.

25 See CTB, *First Annual Report and Statement of Accounts*, 1951.

26 Canada, House of Commons, *Debates*, 30 September 1949: 389–92 (emphasis added).

27 Negotiations had been taking place as early as 1944. See PRO, DO 35/1805, File WQ 351/1, Commonwealth Communications Council Meeting RE: Second Interim Report, 18 May 1944; NA, Records of the Canadian Marconi Company, MG 28, III 72, vol. 22, *Reports and Accounts, 1943–51*; Marconi Papers, vol. 1, N.C. Chapling, Managing Director, Cable and Wireless to J. Findlayson, 31 July 1951; vol. 8, Board of Directors Minutes 1946–58; vol. 19, President's files ..., S.M. Finlayson to Norman Chapling, 20 March 1952.

28 See debate on COTC Act, in House of Commons, *Debates*, 30 September 1949, 389–402, especially the exchanges between Lionel Chevrier, Ross Thatcher, E.G. Hansell of the Progressive Conservatives; see also COTC, *Annual Report, 1969–1970*.

29 PCO, Extract from Cabinet Conclusions Ottawa (hereafter ECC) 13 May 1953, "Trans-Atlantic Telephone Cable; Discussions Concerning Arrangements." See also ECC, Top Secret, Ottawa, 19 May 1953, "Trans-Atlantic Telephone Cable."

30 Ibid., 9 September 1953, "Trans-Atlantic Telephone Cable"; see also Cab. Doc. 174–53, 4 September 1953. The US–UK–Canada cable would have cost the federal government between $3 and $4 million; the Canada–UK cable would have cost at least $15 million.

31 See ibid., 14 October 1953, "Trans-Atlantic Cable; Agreement With UK Post Office and American Telegraph and Telephone Co." See also OIC 1953–1972, 14 October 1953.

32 NA, RG 25, Acc. 1984–85/019, vol. 379, file 10753-40, pt. 1, Telecommunications in Canada Policy general file, Confidential Message from Office of High Commissioner for Canada, London, to SSDEA, 9 November 1954.

33 CTB, *Annual Report 1958*: 7. The proposal was accepted in principle by Commonwealth governments in 1958.

34 House of Commons, *Debates*, 31 July 1956, 6745.

35 PCO, Department of Transport (hereafter DOT), "DOT Submission to the ICTP–Outer Space Radiocommunications," Appendix A. SEACOM was

approved by the Commonwealth Telecommunications Conference held in Malaya in June of 1961. The cable would connect Brisbane, Australia, to Madras to New Guinea, North Borneo, Singapore, Malaya, and Colombo with a spur to Hong Kong. Estimated cost: $68 million. See also CTB, *Annual Report, 1962*. The ICTP was established by cabinet decision on 23 August 1945. The ICTP was reincarnated as the Joint Telecommunications Committee (JTC) at various points. NA, RG 25, vol. 2740, file 6327-40C, pt. 2, meeting of ICTP, 25 January 1950. The departments involved included the DOT, Trade and Commerce, Finance, each of the three armed services, and External Affairs.

36 NA, RG 25, vol. 3740, file 6327-40C, pt. 2, H.O. Moran to George Glazebrook, 16 July 1952.

37 Ibid., George Glazebrook to H.O. Moran, 9 June 1952.

38 Ibid., The International Control of Radio Frequency Assignments, JTC, 2 June 1950; also "The International Control of Radio Frequency Assignments," JTC, 3 May 1950.

39 Ibid., Ad Hoc Committee of the Working Group on Telecommunications Policy, Secret Minutes, 7 April 1955.

40 Ibid., Acc. 1984–1985/019 vol. 379, file 10753-40, pt. 1, Telecommunications in Canada Policy – General file, "Comments on CTB, Report," 6 June 1957.

41 CTB, *16th General Report, 1965, 1967*, 9.

42 NA, RG 33/46, Records of the Royal Commission on Government Organization (RCGO), vol. 295, file 53, H.J. Barrington Nevitt, "Proposed Study of Canadian Government Telecommunications," 10 February 1961.

43 The COTC's 1963 project spending included apportioning $18 million for CANTAT, $11.09 million for BMEWS, $3.5 million for ICECAN, and $27.5 million for COMPAC. The USA–Bermuda cable was apportioned $350,000.

CHAPTER THREE

1 John Zysman, *Political Strategies*, 15.

2 See Hockney and Jesshope, *Parallel Computers*, 3; Flamm, *Creating the Computer*, 13. Flamm has calculated the drop in the price index for computing capacity to be approximately 28 per cent per year, or an order of magnitude every seven years. Flamm, *Targeting the Computer*, 25. Flamm, *Creating the Computer*, 15, 17.

3 K. Flamm, "Technological Advance and Costs," in Crandall and Flamm, eds. *Changing the Rules*, 41.

4 See Kenneth J. Arrow, "Economic Welfare and the Allocation of Resources for Invention," in *The Rate and Direction of Inventive Activity*, 609–25.

5 Nestor E. Terleckyj, "The Growth of the Telecommunications and Computer Industries," in Crandall and Flamm, eds. *Changing the Rules*, 328.

6 Nelles and Armstrong, *Monopoly's Moment*, 325–6.

7 John McDonald, *The Current Status of Electronic Data Processing in Canada*, Study No. 9, "A Skilled Manpower Training Research Programme," (Ottawa: Department of Labour, 1960), 6, 7, 11.

8 J.T. Marshall, "The Philosophy of the Government Committee on Electronic Computers," the Computing and Data Processing Society of Canada, *Proceedings of the Third Conference*, McGill University, Montreal, 11–12 June 1964, 14; see also H.E. Baird, Civil Service Commission, "Justifying Electronic Data Processing in Government Service," the Canadian Conference for Computing and Data Processing, *Proceedings of the First Conference*, University of Toronto, 9–10 June 1958.

9 University of Waterloo Archives (hereafter UWA), file Clippings 0055-1960, Basil Myers, "Education in the Space Age," *IRE Transactions on Education*, September 1960, 66 (emphasis added).

10 See Rowlands, et al., "The Canadian Scene in Computing and Data Processing." Rowlands surveyed twenty-one organizations, seven manufacturing and distributing organizations, six insurance companies, three transportation or utility organizations, four scientific or research type organizations, one military organization and one government department.

11 McDonald, *The Current Status of Electronic Data Processing in Canada*, 23.

12 Research Program on the Training of Skilled Manpower, *A Second Survey of Electronic Data Processing in Canada 1962* Department of Labour Report No. 9C, October 1963, vii; NA, MG 28, I 80, Records of the Canadian Information Processing Society, series 1, vol. 2, C.C. Gotlieb to Michael Starr, Minister of Labour, 9 June 1961; C.C. Gotlieb to J.C. Davidson, Vice-President, Confederation Life, 31 July 1961. Gotlieb cautioned that computing was "too important a matter for the prestige of your department to be weakened by a publication which does not reflect correctly, the consensus of informed opinion in the country."

13 David Thompson, "Data Processing Growing in Canadian Firms," *Financial Post*, 19 May 1962. See also in the same issue, A.D. Cardwell, "Revolution in the Offices."

14 Walter W. Finke, "Truth About Automation: It Keeps the Economy Moving," ibid., 22 September 1962.

15 T.A. Wise, "IBM's $5,000,000,000 Gamble," *Fortune*, September 1966, quoted in Michael Campbell-Kelly, *ICL: A Technical and Business History* (Oxford, 1989), 226.

16 The problem of paperwork was identified not as a technological one, but a human one. Ian Dutton, "Next Problem for Computer: How do

You Speed Up People?" *Financial Post*, 29 September 1962; John H. Aitchinson, "Computer: Key To Savings, Profits," ibid., 18 May 1963; "Datacentres Carry Heavy Loads for Many Companies," ibid., 23 May 1964. By 1964, IBM had set up fourteen datacentres across the country.

17 W.C. Macpherson, "Telephone's Versatility Grows as Machines Talk to Machines," *Financial Post*, 18 May 1963; D.A. Carruthers, "Data Traffic may Eventually 'Out-talk' our Voices," ibid., 23 May 1964.

18 James R. Bright, "Opportunity and Threat in Technological Change," *Canadian Business* 37, May and June 1964.

19 "IBM Technicians Cross Third Frontier," *Financial Post*, 23 May 1964; "Hot Battle to Sell Computers," ibid., 12 September 1964.

20 For a history of ICL, see Campbell-Kelly, ICL. *A Business and Technical History*, 208. See also Hendry, *Innovating for Failure*.

21 UWA, Memorandum to Members of Executive Council from Gayle Johannesen Re: Computing at Waterloo, 1 May 1984. UWA, "Math Faculty at U. of W. To Be First," *Kitchener-Waterloo Record*, 5 April 1967; UWA, clipping file, box 1, file 17, "WATFOR wins world wide welcome," *University of Waterloo Quarterly* 7, no. 3 (November 1966); UWA, 1967 clipping file, box 1, file 9, "Education will be were most computer action is by William G. Davis, Minister of Education, Province of Ontario," *Financial Post*, 3 June 1967.

22 "The burgeoning computer world," *Globe and Mail*, 17 January 1967.

23 NA, RG 20, vol. 2077, file P 8001-7400/E1, R.A. Gordon, "Summary of Relevant Knowledge of IBM Operations," 14 December 1966; R.A. Gordon to C.D. Quarterman, 29 November 1966; R.A. Gordon, "Summary of Branch View of the Computer Situation," 9 November 1966.

24 "IBM Vol. and Variety Grows," *Financial Post*, May 27, 1967; Clive Baxter, "All Computer Men Watch Ottawa," ibid., 27 November 1965. See also "Civil Service Computers at Work to Save Dollars and Improve Bureaucratic Output," ibid., 27 May 1967.

25 NA, RG 20, vol. 2077, file P 8001-7400/E1, R.A. Gordon to C.D. Quaterman, "Meeting with Mr. White of ICT [*sic*], Manager of the Export Division," 29 November 1966.

26 Campbell-Kelley, ICL, 3.

27 NA, RG 20, vol. 1800, Electrical and Electronics Branch, DOI, "An Appreciation Regarding the Market for and Supply of Computers and Associated Equipment in Canada," August 1968; author's interview with Ian Patrick Sharp, October 1993. Mr Sharp was the founder of I. Sharp Associates, a computer service company that began in the mid-1960s.

28 NA, RG 20, vol. 2077, file P 8001-7400/E1, D.B. Mundy to Minister C.M. Drury, "The Relationship of IBM and the Future of the Canadian Data Processing Industry," 30 September 1965.

29 See Lukasiewicz, "A New Role For Canada," 5–6. See also NA, MG 31 J 6, A. Dunn Papers, vol. 22.

30 NA, RG 20, vol. 2077, file P 8001-7400/E1, Memorandum to Deputy Minister, material prepared for briefing prior to meeting with Electronic Industries Association, June 1965.

31 The DBS purchased a 705 Model 3, and the National Research Council bought an IBM 360; interview with Mr George Fierheller, vice chairman, Rogers Cantel Communications, 10 October 1993. Mr Fierheller was IBM's federal government marketing manager in the early 1960s.

32 This is precisely what George Fierheller managed to do in 1969.

33 Pyke, Jr, "Time Shared Computer Systems," in Alt and Rubinoff, eds., *Advances in Computer Systems*, 40.

34 For a thorough description of the technical details of time-sharing, see Dewalt, *Building a Digital Network*, 20–3.

35 Fisher, McKie, and Mancke, *IBM and the U.S. Data Processing Industry*, 317.

36 See Janisch, "The North American Telecommunications Industry from Monopoly to Competition," in Button and Swan, eds. *Regulatory Reform*.

CHAPTER FOUR

1 Hughes, "The Dynamics of Technological Change," in Dosi, Giannetti, and Toninelli, eds. *Technology and Enterprise*, 101.

2 See Janisch, "The North American Telecommunications Industry from Monopoly to Competition," in Button and Swan, eds. *The Age of Regulatory Reform*, 8, 9.

3 Scitovsky, *The Joyless Economy*, 57.

4 Cited in Walker, *Technology, Industry and Man*, 5. See also Ward, *Spaceship Earth*.

5 Grant, *Technology and Empire*, 113 and 129. See also Andrew and Planinc, "Technology and Justice," in Emberley, ed. *By Loving Our Own*, 179.

6 Carey and Quirk, "The Mythos of the Electronic Revolution," 220.

7 Bright, "Opportunity and Threat in Technological Change," *Harvard Business Review*, December 1963.

8 NA, RG 97, vol. 349, file 5200-1, pt. 2, M. Stanley Burke, "Canada: Communications Casualty or World Leader?" no date. One of the recommendations of the Burke report was an 'Electronic' University on Prince Edward Island. See Commons Standing Committee on Transportation and Communications, *Evidence*, No. 9, 6 and 18 November 1969, 45. See also NA, RG 97, vol. 158, Deputy Ministerial Diaries of Allan Gotlieb, "Telesat and the Role of the DOC"; and Eric Kierans, article in *Montreal Star*, 9 September 1969.

262 Notes to pages 93–8

9 Bickers, "The Politics of Regulatory Design," 46. See also Skowronek, *Building a New American State.*

10 NA, RG 33/46, vol. 296, file 1, Royal Commission on Government Organization (RCGO), Project 7 Communications: Initial Report to the Commissioners, part 1, chapter 6.

11 Ibid., file 14, Bell Canada, "Brief to the RCGO, Project 7: Communications," August 1961.

12 RCGO, Initial Report to the Commissioners, pt. 1, chapters 1, 3, and 5; and NA, RG 33/46, vol. 293, file 38, C.M. Brant to F.G. Nixon and Privy Council Telecommunication Policy Committee, 8 January 1961.

13 Hans von Baeyer noted that the RCGO's views on policy "definitely upset Mr. Nixon who claimed that the Government already had a policy." See NA, RG 33/46, vol. 293, file 38, Von Baeyer, Project 7 Interview Report, DOT, TEB, 18 October 1961.

14 Ibid., vol. 295, file 53, B. Nevitt, Proposed Study.

15 NA, RG 25, vol. 248, file 10753-40, pt. 2, Acc. 1990-91/001, DOT Memorandum on Domestic Transcontinental Communications Systems. The estimated cost of the project was put at of $40 million to $60 million; see Douglas Harkness, in House of Commons, *Debates*, 15 March 1962. Indeed, over $11 million was being spent by the COTC to provide the Cape Dyer–Deer Lake coaxial cable link that formed part of the eastern section of the Ballistic Early Warning Systems (BMEWS). The network was administered by CPT from Montreal to Melville, Saskatchewan, with CNT responsible for the Melville to Vancouver portion. Canadian Pacific Archives (hereafter CPA) general FILES, "History Of Canadian Pacific Telecommunications," January 1963, 2. The private wire service provided a communications facility for exclusive use. In 1964 there were over four hundred such dedicated lines in Canada.

16 See Doern and Brothers, "Telesat Canada," in Tupper and Doern, eds., *Public Corporations and Public Policy.*

17 NA, RG 97, vol. 69, file 2, R.K. Brown, "Summary of Canadian Space Activities," November 1968. Atmospheric research had been conducted since the Second World War by the Defence Research Board. See Surtees, *Pa Bell*, 153. The Alouette I was actually a "topside sounder" device.

18 PCO, confidential, DOT, "Memorandum for the Interdepartmental Committee on Telecommunications Policy. U.S. Proposal for Integration of Commercial and U.S. Government (primarily military) Satellite Communications Systems," 13 March 1964, 1.

19 PCO "Memorandum for the ICTP," 13 March 1964. See also NA, RG 19, vol. 4964, file 4181-04-3, E.F. Gaskell to C.G.E. Steele, Secretary to the Treasury Board, 16 March 1964. Participating countries were asked to commit their share of 15 per cent initially, to be followed by further

subscriptions of 30 per cent in 1965, 45 per cent in 1966 and a final 10 per cent in 1967 to be followed by an agreement to open the shares up to other countries.

20 PCO, "Memorandum ...," 9; NA, RG 97, vol. 188, file M-2, pt. 1, B.A. Walker, "Memorandum to Cabinet, Canadian Space Program," 30 July 1969.

21 Cited in Carey and Quirk, "The Mythos of the Electronic Revolution," 414.

22 NA, RG 97, vol. 190, file T-3, pt. 1, "Report of the Interdepartmental Telecommunications Committee," Gilles Sicotte, 1 November 1965. In 1966 GTE, an American company which owned BC Tel, also sought control over Quebec Telephone and the Okanagan Telephone Company, raising its control of the Canadian telephone industry to 11 per cent.

23 NA, RG 19, vol. 4964, file 4181, file Pocket Secret, "Report of the Task Force on Domestic Satellite Communications," October 1967.

24 White House Message on Communications Policy Address to the Congress of the United States by President Lyndon B. Johnson, 14 August 1967; NA, RG 19, vol. 4964, file 4181-08, Annex 2, Appendix D, file Pocket, "Extracts from a Discussion of Studies of Domestic Satellite Communications for the United States, Prepared by the Director of Telecommunications Management, Executive Office of the President, July 1967."

25 NA, RG 97, vol. 69, file 4, "Memorandum from PCO Re: General De Gaulle's Visit," 19 July 1967. Discussions with France were paralleled with the attendance of the vice-president of Radio-Québec, Jacques Gauthier, at Intelsat's meeting in March 1968.

26 Ibid., vol. 187, file I-1, Reynolds to Langley, 26 November 1968; see also "Memorandum to Cabinet re: Definitive Arrangements for Intelsat Conference," 29 January 1970.

27 Claude Beauchamp, "Un Satellite, Au Service du Québec sillonnera l'espace dans 4 ou 6 ans," *Le Soleil*, 1969, n.d.; "Compétences du Québec et rélations internationales," *Le Devoir*, 2 juin 1967.

28 NA, RG 97, vol. 69, file 4, telegram from Canadian Embassy to DEA, Ottawa, 17 November 1967, RE: Cooperation Franco-Québécois, Affaires Exterieures 3532; ibid., vol. 355, file 5300-50/5, pt. 3, A.E. Gotlieb to R. Robertson, 17 December 1970.

29 Ibid., vol. 69, file 4, J.H. Chapman to D.H.W. Kirkwood, 20 September 1967.

30 J. Gauthier, "Et le Québec ... La Juridiction du Satellite, Un Objet de Conflit," in *Le Soleil*, 17 février 1968; "Front Uni Entre le Québec et le Canada," *La Presse*, 29 May 1969.

31 "The Satellite, An Answer to Our Needs," remarks by the Hon. Eric Kierans, at "Rencontre 69" held by the Conseil des Sociétés

Canadiennes-Francaises du Sud de l'Ontario," 7 June 1969; and "A Humanism to be Created: Remarks to the Canadian Consumer Loan Association 25 June 1969 by the Hon. Eric Kierans."

32 NA, RG 97, vol. 69, file 18, Annex to Treasury Board Submission, Government Reorganization 1968, Communications Department, 4.

33 NA, RG 97 vol. 69, file 7, "Operations (Including Regulations) Terms of Reference," 20 December 1968; ibid., file 18, 6th Meeting of Task Force 4, DOC, L.G. Crutchlow, chairman's notes, 31 October 1968.

34 NA, RG 97, vol. 186, file D-17, Richard Gwyn to Allan Gotlieb, 25 August 1970.

35 Ibid., Gwyn to Gotlieb, 29 January 1971.

36 See PCO, C.D. 543/69, Confirmation of the Decisions of Cabinet Committee, 26 May 1969.

37 NA, RG 97, vol. 190, file T-3, pt. 1, Memorandum to Cabinet, Proposed Telecommunications Mission, 31 March 1969; Gotlieb, Telecommission Profile of Work Program, 3 July 1969.

38 Bell Canada Archives Regulatory Information Bank (hereafter RIB), "National Policy on Telecommunications Strategic Concepts for Bell," n.d. (likely 1968), 1.

39 Ibid., notes of J.A. Harvey; notes of meeting in O. Tropea's office on 29 July 1968 between O. Tropea, P.E. Skelton, A.J. De Grandpre, J.A. Harvey, R. Bushfield and G. Haase; "Service Objectives," 1969.

40 Ibid., "Service Objectives," 1969; T. Ringereide to J.A. Harvey, Subject: Government Task Force on National Telecommunication Policy and Legislation, 6 March 1969, 1-2.

41 Ibid., "National Policy on Telecommunications Strategic Concepts for Bell," A.3. Wired City Concept, 1-2.

42 Ibid., B.2, Regulatory and Planning Bodies, 5.

43 NA, RG 97, vol. 70, file 32, "Task Force on Satellites Interim Report on Domestic Satellite Communications."

44 Commons Standing Committee on T&C, Evidence 9 December 1968, 166.

45 The Niagara Power proposal would have created a public company called the Canadian Satellite Corporation (CANSAT). NA, RG 97, vol. 71, file 35, Canadian Embassy in Washington to SSDEA, 28 April 1967; Memorandum from Interdepartmental Telecommunications Policy Committee/USSDEA to Canadian Embassy Washington, 20 March 1967. In July 1967 the government set up a task force on satellites under the aegis of the Privy Council Office. Four months later, the task force issued its report which was subsequently amplified into a white paper entitled "A Domestic Satellite Communications System for Canada."

46 PCO, C.D. 751/70 Procurement of Satellites and Financing of Telesat Canada, 17 June 1970. See also "Uniting Canada from 22,300 mi. up," *Ottawa Journal*, 4 April 1968.

47 The DOC was even contemplating the idea of refusing microwave facilities licenses in the hopes that it would increase the use of the satellite.

48 Doern and Brothers, "Telesat Canada," in Tupper and Doern, eds.,
Public Corporations, 244.

49 The cabinet also ordered RCA to let contracts to Northern Electric on
the communications transponder system.

50 House of Commons, *Debates*, 14 April 1969: 7503; Commons Standing
Committee on Broadcasting, Film and Assistance to the Arts, *Evidence*,
6 March 1969.

51 Commons Standing Committee on T&C, Evidence 11 March 1971, 11.

52 *White Paper: A Domestic Satellite Communications System For Canada*
(Ottawa, 1967).

53 The federal government's experience with ground segment began in
the mid-1960s with the construction of the Mill Village ground station
for the COTC. See NA, RG 20, vol. 2043, file 8001-5895/51-1, "Memorandum: Canada's Electronics Industry and Telecommunications Systems," J.M. Mercier, 15 January 1969.

54 Ibid., Léon Balcer to Eric Kierans, 12 August 1970.

55 Author's interview with Kierans, 4 June 1992. Peter Ward, "A Pitch by
'neutral' Kierans'," *Toronto Telegram*, 17 July 1970.

56 PCO, C.D. 953-69, Confidential, Memorandum to the Cabinet. A Communications Technology Satellite for Canada, 25 September 1969.

CHAPTER FIVE

1 BCA, RIB, Bell Telephone Company of Canada (hereafter BTCC) President's Conference, 14–16 January 1965 at the Seigniory Club, PQ, W.H.
Cruickshank, "Public Relations," 3, 5.

2 Royal Commission on Transportation, *Report, Vol. II* (Ottawa, 1961), 160.

3 Commons Standing Committee on Transport and Communications
(T&C), *Evidence*, No. 42, 14 March 1967.

4 Ibid., *Evidence*, No. 4, 31 October 1967, Appendix A, Industrial Wire &
Cable Co. Ltd., 150.

5 *Evidence*, No. 155; "How effective is government regulation?" IW&C
advertisement from the *Montreal Gazette*, 18 May 1967; *Evidence*, No. 6,
16 November 1967, 217.

6 NA, RG 97, vol. 78, file 6900-7-14, D.S. Marshall, *DCF Systems Limited
Report on Current Operating Methods of Bell* 29 September 1967; *Evidence*,
Testimony of M.V. Holt, 341.

7 An Act respecting the BTCC, 16–17 Eliz. II, c. 48, assented to, 7 March
1968. The section referred to here is section 6, which repealed section 5
of c. 81 of the 1948 act. House of Commons, *Debates*, 15 February 1968,
6771.

8 NA, RG 46-cII-I, vol. 1679, file 49404, J.W. Pickersgill to Minister of Transport Re: Regulation of Telecommunications, 2 August 1968.

9 RIB, BTCC, *Submission to the CTC*, *"Rate of Return on Capital"*; A.J. de Grandpre to C.W. Rump, Secretary, Railway Transport Committee, 27 December 1967."Ruling on Bell Earnings Limit Seen Possible in Few Weeks," *Montreal Gazette*, 25 January 1968; "Won't Force Bell Telephone To Cut Rates 'At this Time'," *Ottawa Journal*, 24 January 1968; "Ontario urged to force down telephone rates," *Globe and Mail*, 30 January 1968; "Bell Telephone Company made millions 'illegally,' NDP says," *Toronto Daily Star*, 1 February 1968.

10 RIB, Bell Canada, *Memorandum on Corporate Objectives*, Exhibit B-161, 1969 Rate Case, 5–7.

11 CTCT, file C 955.178, Testimony of R. Scrivener, vol. 40, 171, 199; also Kyles, direct examination, vol. 54, 9 June 1969, 2171.

12 RIB, CFMM, in Answer to the Application of Bell Canada 16 January 1969. CTCT, vol. 80, 30 July 1969, 6261–5, Carroll in Argument.

13 See Rens, *L'Empire invisible*, 269ff, where he quotes John Kettle, *The Kettle Text*, 45–61. The passage recounts the reaction to the 1969 decision in Bell's senior management cadre. BCA, file: Regulation General, 1970–1979, Pitfield, Mackay, Ross & Company Special Report, "The Canadian Telephone Industry Part I – Regulation and the Telecommunications Industry," October 1971.

14 CTCT, vol. R9K-70, file C 955.180, "In the Matter of An Application of Bell Canada ..."; RIB, Memorandum on Construction Program June 1970.

15 BCA, IB, 2 December 1970. The construction program was reduced by 12 per cent, and Bell stocks were downgraded from high grade to upper medium grade. NA, RG 97, Acc. 1989–1990/218, box 6, file C 5608-4, pt. 2, Bell Canada, "Bell Canada 1971 Rate Application Management Background Information."

16 CTCT, Application of Bell Canada ... 1971; Canadian Radio-Television and Telecommunications Commission Archives (hereafter CRTCA), Public Archives Records Centre (hereafter PARC), box 135, CTCT, BCRA, file C 955.181, vol. 5 R-2/72, 1 March 1972.

17 House of Commons, *Debates*, 6 October 1971, 8483; 6 November 1971, 9402; 9 November 1971, 9467–8.

18 Editorial, *Toronto Sun*, 10 November 1971.

19 CRTCA, PARC, box 135, CTCT, BCRA, 1 March 1972, file C 955.181, vol. 5a, Memorandum of Evidence of W.A. Allan Beckett on Economic Review, 189F-189CC; box 136, CTC 23, R-2/72, 27 March 1972, Argument by Bell counsel J.S. O'Brien, 2873; Argument of W.B. Williston, 2955, 2959; 28 March 1972, Argument of D.W. Burtnick, 3100–2; Argument of B. Lesage, 3011, 3019, 3038.

20 Reported in 1972 CTC, Bell Canada file C-955.181, Telecommunications Committee, 19 May 1972.

21 Montreal's telephone count, for example, increased 64.1 per cent between 1958 and 1971, and Toronto's by 43.7 per cent. Yet subscribers in both cities paid only between 3 and 8 per cent more.

22 NA, RG 97, Acc. 1989–1990/218, box 6, file C 5608-4, pt. 2, Memorandum 14 April 1972 A.E.G. To file Re: Meeting Between Mr. Stanbury and Mr. Scrivener, 12 April 1972.

23 Ibid., A.E. Gotlieb to the Minister Re: BCRA filed with CTC, 10 November 1972.

24 House of Commons, *Debates*, 16 January 1973, 374 (Terry Grier); 5 February 1973, 988–9 (Perrin Beatty).

25 NA, RG 97, Acc. 1989–1990/218, box 6, file C 5608-4, J.W. Halina Memorandum Re: Bell Rate Application, 29 November 1972.

26 Ibid., A.E. Gotlieb to Minister Re: Representation of the Individual in the Rate Hearing Process, December 1972.

27 Ibid., A.E. Gotlieb to Minister Re: to Mr. Benson of CTC, December 1972.

28 1973 CTC Cases, Bell Canada Application "A," file C-955.172, Telecommunications Committee: Hon. E.J. Benson, President, and Commissioners Frank Lafferty and Anne H. Carver, 30 March 1973.

29 Canadian Consumer Council, *Annual Report 1971* (Ottawa, 1971), 33, 34.

30 RIB, CCC, "Report on the Consumer Interest in Regulatory Boards and Agencies," 1973, 5.

31 Ken Rubin, "A Glimpse at the National Citizen Participation Power Structure and Your Reaction," n.p., September 1972.

32 RIB, Maryon Brechin, CAC, "Talk to the Canadian Telecommunications Carriers Association 1973."

33 OA, RG 14 MTC, Div. Subject Correspondence FILES (hereafter DSCF) 1975, box 10, file 5100-1-25, Common Carrier Regulatory Policy. See also Michael J. Trebilcock, "Winners and Losers in the Modern Regulatory States: Must the Consumer Always Lose?" n.p., 1975.

34 CRTCA, PARC, box 140, CTCT, BCRA file C 955.182, vol. 23 T-1/73, 13 February 1973, Argument by Mr. O. Platt, Counsel, Native Marathon Dreams, 2466–9.

35 CRTCA, 1974 CTC Exhibits, Individual Information Institute, Submission to the Canada Transport Commission, 10.

36 CRTCA, PARC, box 140, CTCT, BCRA file C 955.182, vol. 24 T-1/73, February 1973, 2754.

37 NA, RG 97, Acc. 1989–1990/218 box 6, file C 5608-4, pt. 2, Memorandum Ken Wyman to Deputy Minister Re: Ontario's Stance at Recent Bell Rate Case, 15 March 1973.

38 CRTCA, PARC, box 140 CTCT, BCRA file C 955.182, vol. 26 T-1/73, 16 February 1973, 3007; and 25 January 1973, 1225, 1227. Bell planned to devote $96 million to innovation and modernization programs.

39 NA, RG 97, Acc. 1989–1990/218, box 6, file C 5608-4, pt. 2, Memorandum F. Bigham to K. Wyman Re: A Review of Discussions within the present Bell Rates Cases Re: Possible construction cutbacks, February 1973.

40 House of Commons, *Debates*, 2 April 1973, MPS David Lewis (York South), and Arnold Peters (Timiskaming), 2890–1.

41 Ibid., Hon. Alvin Hamilton (Qu'Appelle-Moose Mountain), 2883–4; Mark MacGuigan, 2885; Gérard Pélletier, 2882–3; ibid., 6 April 1973, 3036.

42 PCO, Cabinet Minutes (Secret), Serial no. 17–73, Thursday, 5 April 1973.

43 PCO, Memorandum to the Cabinet Communications: Representation of Consumer Interests in the Telecommunications Regulatory Process Before the CTC, 27 March 1973.

44 Other alternatives were to have a separate office outside the commission but within either the DOC or the DCCA; to have a separate office within the government but outside departmental structure, like the Bureau of Pension Advocates; to establish an ombudsman; or to create a separate office within the regulatory commission operating on a full-time basis. None of these alternatives met with complete approval.

45 PCO, Cab. Doc. 446–73, 9 May 1973. The cabinet had approved the establishment of a single regulatory agency in July 1972. See Cab. Doc. 780/72 and Cabinet Decision of March 1973 approving publication and tabling "Proposals for a Communications Policy for Canada."

46 PCO, Memorandum to the Cabinet, "Interim Reform of the Federal Regulatory Structure – Communications," 16 May 1973.

47 PCO, Cabinet Minutes, 31 May 1973.

48 CRTCA, PARC, box 146, CTCT, BCRA file C 955.182.1T-2/74, vol. 1, 4 February 1974, V. Dowie and K. Rubin.

49 Ibid., box 148, CTCT, BCRA file C 955.182.1T-2/74, vol. 42, argument by E. Saunders, 5998, 6010.

50 Ibid., 29 May, 1974.

51 Ibid., 30 May 1974, argument by D.W. Burtnick; NA, RG 46 C II-1, vol. 1681, file 49645.1, pt. 9, Province of Ontario, Direct Evidence Rate Making and the Regulatory Process, Bell Canada Application B, 29 March 1974, amended 29 April 1974; RIB, Consumers' Association of Canada, *Amended Application "B" Memoranda of Evidence*, 29 March 1974.

52 CRTCA, PARC box 148, R. Cohen Canter for Public Interest Law to CTC, "The CTC Rings Bell's Chimes," 16 July 1973.

53 1974 CTC Cases, Bell "A," 459; Anthony J. Patterson, "Approval of Bell rates reveals wider implications," *Ottawa Citizen*, 16 August 1974.

54 Letter from Bert McNeill, Stratford and District Labour Council, *Stratford Herald*, 28 August 1974; Laurent Laplante, editorial, *Le Devoir*, reprinted in *Guelph Mercury*, 27 August 1974.

55 OA, RG 14, MTC Division Subject Correspondence, files Box T-203, Communications Branch Activity Report, W.A. Rathbun to A.T.C. McNab, Deputy Minister, 29 October 1973.

56 Ibid., MTCDSCF, box T-210, file Ctees: Ontario Chamber of Commerce. W.A. Rathbun to Mr. Inglis of Ontario Chamber of Commerce, 20 November 1972; ibid., file Select Ctee. on Communications, Gordon Carton et al., Outline Address by the Hon. G. Carton, Minister of Transportation and Communications during Budget Debate, 25 May 1973.

57 Ibid., box T-213, file Boards – Policy and Priorities Board, A.T.C. McNab, "Presentation for Policy and Priorities Board of Cabinet by Communications Branch Policy Development Division, MTC," 20 January 1972.

58 Ibid., 1974, box TB-12, FILE 5100-1-32, Ron G. Atkey to W.A. Rathbun Re: Summary of Discussions with Officials of Your Branch, 27 August 1974.

59 Ibid., 1975, box 14, file 5103-1-3, Bell Canada General, J.A.D. Ham to R. Bulger Re: Bell Canada, meeting 19 March 1975.

60 Ibid., 1977, box TB-15, file 5103-11-5, Bell Canada Rate Case, W. Darcy Mckeough to Premier W. Davis, 21 February 1977. Although this document was written later, it accurately imparts the essence of the debate between the moderates and the forward party within the Ontario government.

61 See Ian Rodger, "Bell endorses CTC cost formula for calculating telephone rates rises," *Globe and Mail*, 4 March 1975

62 OA, RG 14, MTCDSCF 1974, box TB-04, Rhodes to Pélletier 28 October 1974; G. Pélletier to J. Rhodes, 22 November 1974; Rhodes to Pélletier 28 October 1974.

63 Ibid., 1975, box 2, Andrew J. Roman, general counsel, Submission of the Consumers' Association of Canada to the CTC Telecommunications Committee, "In the Matter of an Automatic Rate Adjustment Formula for Federally Regulated Telecommunications Carriers," 5 March 1975; box TB-08, Briefing Materials to Minister, "Relationship of Ontario to the Federal Government and its Regulatory Bodies."

64 RIB, J.P. L'Allier to G. Pélletier, 27 November 1974.

65 CRTCA, PARC, box 166, CTCT, "In the Matter of an application of Bell Canada dated May 30, 1975," file C 955.183. This hearing was known as hearing "A." The hearings on the interim application were designated "B."

66 BCA, Standby Information Subject: Comment on CTC Decision of 28 July, 29 July 1975, Statement by A.J. de Grandpre.

67 Geoffrey Stevens, "The Bell and the CTC (I)," *Globe and Mail*, 28 August 1975.

68 Geoffrey Stevens, "A boodle for Bell," *Globe and Mail*, 27 December 1975.

CHAPTER SIX

1 This term is frequently used to describe the period of innovation of the 1960s and into the 1970s. In its historical usage, it denoted a vague series of technical changes in transmission and storage of information, extension of telecommunications systems by satellite and microwave, and so on. Cf. Landes, *The Unbound Prometheus*.

2 "Nerve system of the nation, communications belong to the public," *Electronics and Communications*, December 1968 (n.p). See also NA, RG 97, vol. 185, file D-2; also Montreal *Star*, 9 September 1969.

3 NA, RG 97, vol. 78, file 6900-6-4c-9, McGee to N. Phemister, 1 December 1970.

4 Ibid., vol. 111, file 6901-6-6-2, G.E. Inns to R. Gwyn, 26 November 1969; Saskatchewan Archives (hereafter SA), Gordon Grant Papers, Collection 45, file 291a, Minutes of Special Meeting of Saskatchewan Telecommunications, 28 May 1971.

5 Ibid., box 78, file 6900-10-1, "Canadian Communications / Outlook and Alternatives," speech by Eric Kierans, 11 February 1971; "Communications: Ottawa Cherche a Associer l'Interet National et Progrès des Societés Privés," *La Presse*, 29 septembre 1970.

6 NA, RG 97, vol. 100, file 1220-5, pt. 7, DOC, Draft Memorandum to Cabinet, 16 March 1971; P.E. Trudeau to Gérard Pélletier, 23 juillet 1971.

7 Ibid., vol. 359, file 5400-0, pt. 2, G. Pélletier, Project de discours sur "une politique culturelle nationale," 17 November 1969.

8 McRoberts and Posgate, *Quebec: Social Change and Political Crisis*, 96.

9 NA, RG 97, 117, box 1, Acc. 1984–85/507, "The Case for a Quebec Policy on Communications," May 1971, Ministère des Communications du Québec, 59–60.

10 Ibid., 9.

11 See Assemblé Nationale, *Journal Des Débats*, Commission permanente de l'éducation, des affaires culturelles et des communications, 15 mai 1973, B-1567; Ian Rodger, "Quebec Has Two Ears but No Mouth," *Financial Post*, 27 November 1971.

12 Province of Ontario, *Communications Co-ordinating Committee Report to Cabinet Committee on Policy Development in the Matter of a Telecommunications Policy*, 18 February 1971, 3, 7; OA, RG 14, box T-209, file 18/23/4/6, "Highlights of Communications Coordinating Committee. Report to Cabinet," 25 January 1970.

13 OA, RG 14, box T-202, file MTC, Organization Background, "Remarks by W.A. Rathbun to the Departmental Series of Divisional Presentations," 9 February 1972.

14 Another element of Ontario's burgeoning interest may have been the DOC's freeze on the issuance of microwave licences to Ontario Hydro, a major point of conflict. See OA, RG 14, box T-212, Mid-Ontario Microwave Network, R.P. Bulger, manager Common Carriers Section, to W.A. Rathbun, 19 November 1973.

15 Ontario Legislature, *Debates*, 30 March 1971, 5; author's interview with William A. Rathbun, former director of Communications Branch, Ministry of Transportation and Communications, 22 May 1992.

16 OA, RG 14, box TB-06, file 1110-1-6 61, Government of Ontario Working Paper – Communications Policy Objectives, February 1974.

17 Ontario Legislature, *Debates*, 25 May 1973, 2291.

18 NA, RG 97, vol. 177, "Notes for a Speech by Hon. G. Pélletier, in Debate on Speech from the Throne," 17 January 1973.

19 PCO, Memorandum for Members of the Interdepartmental Committee on Communications, "Communications and the Future of Canada," prepared by Henry Hindley, 16 January 1973.

20 NA, RG 97, vol. 365, file 1020-12, pt. 3, Jules Léger to Michael Pitfield, Secretary to the Cabinet (Plans), 9 January 1973; Pierre Juneau to Allan Gotlieb, Deputy Minister, DOC, 12 January 1973.

21 Minister of Communications, *Proposals for a Communications Policy for Canada: A Position Paper of the Government of Canada*" (Ottawa, 1973), 7, 23.

22 Commons Standing Committee on Transport and Communications, *Evidence*, No 10, 29 May 1973, 1–10.

23 OA, RG 14, DSCF,1971–73, Commissions file: Federal Provincial Meeting at Ottawa, 5, 6 July 1973 "Meeting of Federal and Provincial Officials" (note prepared by W. Bielski, 13 July 1973).

24 NA, RG 97, Accession 1990–91/142, box 1, file 5050-6, pt. 1 ATB, Memorandum "Discussion with officials of the Four Western Provinces Winnipeg," 11 and 17 January 1973.

25 G. Hutchinson, "Decisions on telecommunications policy ahead – if provinces agree," *Financial Post*, 24 November 1973.

26 W. Johnson, "Green Paper Unveiled," *Globe and Mail* 23 March 1973.

27 André Beliveau, "L'Allier ira jusqu'au bout," *La Presse*, 11 septembre 1973; William Johnson, "Ottawa vs. the Provinces: the communications battle is brewing," *Globe and Mail*, 29 November 1973.

28 NA, RG 46, C-II-1, vol. 1706, file 49645.1, Joint Provincial Statement at Federal Provincial Conference on Communications, 29–30 November 1973.

29 "Provinces Attack Ottawa's Control of Communications," *Globe and Mail*, 20 November 1973.

30 OA RG 14, DSCF, 1974, box TB-06, file: Provincial Officials Meeting 1110–1-6, "Government of Ontario Working Paper Communications Policy Objectives," February 1974.

31 Ibid., box 20, file 5105-1-12, Ontario Telecommunications Policy, J. Rhodes to Hon. William Davis, n.d. 1974.

32 Ibid., box TB-11, file: Federal DOC–General 5100–1, Communications Branch, MTC Report, "Meeting with Mr Pelletier April 9, 1974"; "Bitter Charges Against Pelletier; L'Allier Hardens His Positions," *La Presse* 7 March 1974.

33 OA, RG 14, DSCF,1974 box TB-07, file; Provincial Ministers' Conference 1121–1–10, J. Wouters, Senior Advisor (TEIGA) to G.S Posen re: Meeting of Provincial Ministers of Communications in Victoria, 27–28 May 1974.

34 Ibid., Robert Strachan to Gérard Pélletier, 30 July 1974.

35 BCA, file Regulation: Interconnection of Equipment "Background," 1972.

36 RIB, "PEAS," confidential, 22 March 1973. PEAS was the code word for Bell's interconnection strategy paper. Translated to French, PEAS became POIS – Postures on Interconnection of Systems.

37 NA, RG 97, Acc. 1986–1987/324, file 8916-5, pt. 2, Computer Communications Group of Bell Canada, "Comments by Bell Canada on the 39 Recommendations Contained in 'Branching Out,' November 1972; ibid., box 418, file 7500-1445-1, Electronic Industries Association of Canada, "Brief on the EIAC Reaction to the Government's Position Statement," 1973.

38 I. Rodger, "Computer Firms Complain about Bell Canada's Service," *Financial Post*, 27 September 1969; "Bell Turns a New Leaf to 1970s with more 'truly helpful' service," ibid., 10 January 1970.

39 OA, RG 14, DSCF 1971–1973, box T-220, file: Interconnection, International Business Machines Canada, "Comments on Connection of Customer-Provided Equipment to the Public Switched Network," Confidential, IBM Canada, 16 June 1972; W. Walton, General Manager, Harding Corporation, to Members of Parliament re: Bell Canada, n.d. 1973.

40 Ibid., Federal DOC–Interconnection, J.M. Hillier and "Friend," "Debate on Interconnection and the Canadian Industrial Communications Assembly" at the annual conference in Montreal, 16 March 1972.

41 NA, RG 55, Acc. 1986–1987/379, box 32, file 7148-02(1), "Summary of Policy Statement Appendix B Confidential Draft" (Computer/Communications Federal Policy Statement), January 1973; NA, RG 97, vol. 177, file 4, "Notes for Remarks by Gérard Pélletier, Minister of Communications to the Canadian Industrial Communications Assembly, 15 March 1973."

42 G. Hutchison, "Provinces Tangle over Phone Interconnection," and Hugh McIntryre, "The little guys are snapping at Ma Bell," *Financial Post*, 11 August 1973.

43 NA, RG 97, Acc. 1990–1991/142, box 1, file 5050-6, pt. 1, Memorandum from R. Bulger (n.d.); OA, RG 14, DSCF 1971–1973, Federal DOC – Interconnection, box T-216, file: Interconnection No. 3, October 1973; box TB-15, John R. Rhodes to Mr Robinson, President of Tele-Connect Publications, 13 October 1974; and W. McDougall, "Interconnection," 16 July 1974.

44 OA, RG 14, DSCF 1971–1973, Commissions – Min. Comm. file, Prairie Provinces Communications Group, "A Western Perspective on Communications," November, 1972.

45 Ibid., DSCF 1974, box 13, file: Interconnection Adv. Group Meeting 5102–1–5, R. Bulger to Jim Crowson, DOC, 25 February 1975.

46 Ibid., DSCF 1975, box 12, file: Interconnection 5102–1–4, May 1975, R.P. Bulger, "Policy Speech Canadian Interconnection Magazine," May 1975.

47 ANQ, Côte de Fonds E 10, Versement 87–11–009, Article 13, file Interconnexion, Juin 79 à 80, "CNCP Bell: Une Mauvaise Décision Prise Au Mauvais Endroit," Communiqué de presse, le 6 août 1979.

48 OA, RG 14 DSCF, 1974, box TB-07, file 1121-1-12, Federal-Provincial Conference, M.H. Larratt-Smith to Minister re: "Proposal for a Rhodes/ Pélletier/L'Allier Meeting," 7 November 1974.

49 Ibid., box TB-11, file 5100-1, vol. II, Fed DOC General, W.A. Rathbun to David Hobbs RE: Federal-Provincial Arrangements, 13 November 1974.

50 NA, RG 97, vol. 100, file 1220-5, pt. 7, G. Pélletier to Chairman of Conference of Provincial Ministers Responsible for Communications, 18 September 1974; Commons Standing Committee on Transport and Communications, *Evidence* 1974 "Respecting Estimates, DOC, 1974–1975," 18 October 1974.

51 Gérard Pélletier, *Communications: Some Federal Proposals*, 4.

52 NA, RG 97, vol. 177, Federal/Provincial Conference Press Reports, 12 May 1975; Pierre Bellemare, "L'Allier joue ses dernieres cartes," *Le Soleil*, 15 juillet 1975.

53 William Johnson, "Council of Communications Ministers – Quebec Stays Out. Ontario joins with Reservations," *Globe and Mail*, 17 July 1975; "Communications- Le Québec s'est retrouvé seul," *Le Jour*, 17 juillet 1975; "La fin du Mythe de la Souveraineté Culturelle," ibid., 18 juillet 1975.

54 William Johnson, "Pélletier balks at provincial proposals for transferring communications power," *Globe and Mail*, 12 May 1975.

CHAPTER SEVEN

1 Bickers, "The Politics of Regulatory Design," 46.

2 Stuart Ross, "The Burgeoning Computer World," *Globe and Mail*, 17 January 1967; UWA, clipping files, box 1, file 21, "Victory at Waterloo,"

CIPS Magazine, February 1971, 9. See *A Science Policy For Canada. Report of the Senate Special Committee on Science Policy. Volume 1: A Critical Review: Past and Present* (Ottawa, 1970), 83, where the committee witness Arthur Porter notes that Canada "was the first nation to set up three individual data-handling systems and to tie them together over radio links."

3 Ian Reid, "Profile of the Canadian Computer Industry," *Executive* 10 (December 1968).

4 Ian Rodger, "CNCP Rocking Computer Boat," *Financial Post* 8 March 1969.

5 NA, RG 19, vol. 4963, file 4170-02, pt. 2, Memorandum to the Cabinet, *Amendment to the Railway Act Concerning the Regulation of Telecommunications*, 30 April 1969.

6 Hyman Solomon, "Ottawa: Hands Off Quebec Computers," *Financial Post*, 3 May 1969.

7 *Evidence* No. 9, Respecting Bill C-11 ...," 6 and 18 November 1969.

8 PCO, Memorandum to Cabinet re: Canadian Computer/Communications Agency, 15 September 1970, revised 15 October 1970.

9 NA, RG 55, Acc. 1986–87/259, box 32, file 7148-02(1) FP, "Brief on Public Policy and the Marriage of Communications Services with Computer Services," the Computer Group, 23 July 1969. This computer group was composed of most of the major data processing utilities in the country. Technological balance of payments represents an account of payments foreigners have made to Canada for use of patent techniques or technical expertise, minus Canadian payments for foreign patents and technical expertise.

10 Eric Kierans, "After the Telecommission," speech to the Toronto Branch of the Data Processing Management Association, 19 April 1971.

11 NA, RG 97, vol. 187, file E-14, "Terms of Reference for the Study of the Relationships between Common Carriers, Computers Service Companies and their Information and Data Systems," Fall 1969.

12 Ibid., "Submission to the Government of Canada, Department of Communications – Study of the Relationships Between Common Carriers, Computer Service Companies and their Information and Data Systems," December 1969.

13 NA, RG 97, vol. 78, file 6900-10-1, Eric Kierans, "The Computer: Some Canadian Options," text of a speech given before the Montreal Chapter of the Data Processing Management Association, 9 February 1970.

14 NA, RG 97, vol. 78, file 6900–10-1, Eric Kierans, "Canada and the Computer Utility," text of a speech, 14 January 1970; ibid., Kierans, "The Computer: Social and Economic Aspects," text of a speech, 23 February 1970.

15 Ibid., Acc. 1989–90/185, box 13, file 5635-1, pt. 1, Memorandum to the Cabinet, "Participation by Telecommunications Carriers in Public Data-Processing," 23 February 1970.

16 Ibid.

17 McGoldrick's *Handbook of the Canadian Customs Tariff and Excise Duties* (Montreal: McMullin Publishers, annual) reported that under Import Classification 414c, the British preferential tariff was free on computers, the most favoured nation rate was 10 per cent, and the general rate was 25 per cent; DBS, External Trade Division, *Trade of Canada, Exports by Commodities 1970*, 880.

18 Kierans and Pépin did note, however, that if there was a "quantum reduction" in the cost of smaller machines such as reasonably powerful free-standing mini-computers, the trend could be towards smaller and more decentralized computing. This is, of course, exactly what happened. They went on to inform cabinet, however, that such technological development might also increase the viability of utility-based computer power. "It is reasonable to assume that the present trend to concentration of data-processing will probably continue for most of the next decade."

19 NA, RG 55, Acc. 1986–1987/359, box 33, file 7148-06, "Annex C to Memorandum to Cabinet, 23 February 1970, Summary of Responses to Question 2 of the Study of the Relationships Between Common Carriers, Computer Service Companies and Their Information and Data Systems," 1.

20 Ibid., 5.

21 Ibid., 7, 4.

22 I.P. Sharp called the relationship between the computer service industry and the common carriers "appalling." "They wrote their own ticket because they were monopolies," he noted, and "the state succeeded in protection by omission the common carriers, and failed to protect the high technology." Author's interview with Ian Sharp, October 1993. For example, the computer services industry complained that the 5 level code the carriers used for data transmission was inappropriate, and should be converted to a unified 8 level code (ASCII) to "enhance the growth of [the] computer utility."

23 NA, RG 97, vol. 158, Deputy Ministerial Diaries, Memorandum to the Minister from A.E. Gotlieb, 24 March 1970; editorial, "Keep the Flow East and West," *Toronto Star*, 18 April 1970.

24 NA, RG 19, vol. 4963, file 4170-02, EDD Telecommunications Policy Development, R.B. Bryce to the Minister, "Cabinet Committee Agenda – Participation by Telecommunications Carriers in Public Data Processing," 9 March 1970, 2–3.

25 Ibid., S.S. Reisman to Acting Minister.

26 NA, RG 55, Acc. 1986–87/359 box 33, file 7148-06, Confidential, S.S. Reisman to C.M. Drury RE: Cabinet Document 245–70, Participation by Telecommunications Carriers in Public Data Processing, 16 March 1970.

27 The OECD's Directorate of Scientific Affairs also recognized the burgeoning importance of the computer utility, giving international support to the idea.

28 NA, RG 97, vol. 100, file 1220-5, pt. 7, A.W. Johnson to C.M. Drury, Cabinet Committee on Economic Policies and Programs, 1 October 1970.

29 NA, RG 19, VOL. 4934, file 3710-02(1), Acc. box. 35 EDD Computer Communications General Series Policy Development, John N. Turner to J.L. Pépin, 23 February 1972; Peter Vivian to Minister RE: CDC Proposal for an Indigenous Computer Industry in Canada, 21 February 1972.

30 NA, RG 97, file 8916-5, pt. 2, Acc. 1986–87/324, Consumers Association of Canada, "Some Notes on the Report of the CCCTF," December 1972: 4, 16; IBID., "Branching Out – Far Enough? A Comment on the Report of the CCCTF by the Vanier Institute of the Family," 9 November 1972.

31 Ibid., "Comments on Branching Out by IBM Canada Ltd.," November 1972; "Comments by Bell Canada on the Thirty-nine Recommendations in *Branching Out: Report of the Computer/Communications Task Force Report*"; see, for example, "Stimulating Development of a National Data Network," which, despite its title, had for its objectives "achieving economies in the development and use of communications networks …"; "development of Canadian expertise in design, manufacturing and service capabilities," and, vaguely to "foster the general development of Canadian computer/communications"; E.J. Payne "Developing a Canadian Computer/Communications Industry," December 1974.

32 Ibid., Electronic Industries Association of Canada, "Brief on the EIAC Reaction to the Government's Position Statement," 1973.

33 Ibid., Canadian Telecommunications Carriers Association, "CTCA Comments on Computer/Communications Policy," November 1973.

34 OA, RG 14 MTC, DSCF, box T-221, file Task Force CCC/TF "CCC/TF Provincial Meeting," 1971.

35 Ibid., box TB-06, file Provincial Officials' Meeting Toronto 1110-1-7, R. Bulger to D. Hobbs, Directory Policy Development Branch Re: Computer Communications, 26 August 1974.

36 NA, RG 97, VOL. 407, file 7500-1031-2-A-4, pt. 1, "Increasing the Canadian Presence in the Computer/Communications Industry," n.d.

37 NA, RG 19, vol. 4934, file 3710-02(1) Acc. box 35, EDD Computer Communications General Series Policy Development, Memorandum RE; Draft memorandum to Cabinet Computer/Communications Policy; T.K. Shoyama to S.S. Reisman, 6 February 1973; S.S. Reisman to A.E. Gotlieb, 6 February 1973.

38 NA, RG 97, Acc. 1985–1985/507, box 001, file 1465-16, B.A. Walker, "Data Communications. Aspects of a Policy on Computer/Communications," 15 January 1972, 1.

39 Ibid., vol. 409, file 7500-1031-2-A-9, pt. 1, H.H. Brune, Canadian Computer/Communications Task Force, "EDP User Survey Report No. 6: The North/South Flow," 6 February 1973. Allan Gotlieb, Charles Dalfen, and Kenneth Katz, "The Transborder Transfer of Information by Communications and Computer Systems: Issues and Approaches to Guiding Principles," *American Journal of International Law* 68, no. 2 (April 1974), 229.

40 NA, RG 19, vol. 4935, file 3714-09-5 Acc. box 35, EDD Computer/ Communications Working Group 9, "Working Paper on the Impact of the North South Flow on the Canadian Computer Service Industry and Users, Confidential," 27 June 1973.

41 NA, RG 55, Acc. 1986–87/359, box 32, file 7148-02 (1), FP, Cabinet Memorandum on Computer/Communications Policy, 29 February 1973.

42 Ibid., I.A. Steward and J.C. Madden to N.D. Brewer, Secretary Interdepartmental Advisory Committee on Computer/Communications, 9 February 1973.

43 NA, RG 97, vol. 157, file 7500-1031-2-A-6, pt. 1, K. Thompson, SaskTel, to Dr. H.J von Baeyer, 10 July 1973.

44 OA, RG 14, MTC Division Subject Correspondence files 1971–1973, box T-213, loc. 18/23/6/8, James A. Shantora, Notes on Computer Panel Meeting, 5 May 1970.

45 Ibid., file Computer Utility, J. Shantora, "Activities Pertaining to Computer Utility," December 1970.

46 Ibid., C. MacNaughton to G. Pélletier, 26 May 1971.

47 OA, RG 14, DSCF 1971–1973, box T-213, file Computer/Communications, Communications Branch, "Policy Submission to the Minister of Transportation and Communications on Data Communications and Computer Utilities," 11 February 1972.

48 OA, RG 14, DSCF, box TB-06, file: Prov. Officials' Meeting Toronto, "Computer Communications Issues, Interests and Areas of Concern As Expressed by Various Ontario Government Ministries," 1974.

49 ANQ, Côte de Fonds (CF) 10, versement 88-06-004, article 37, file Téléinformatique, "L'importance de la téléinformatiqe au Québec," 24 septembre 1973; L. Philippe Milot, Memoire, "La Téléinformatique au Québec," 20 Septembre 1973.

50 Ibid., Comité Interministériel sur la Téléinformatique, "Rapport Interimaire sur la Téléinformatique au Québec," janvier 1975.

51 NA, RG 97, vol. 408, file 7500-1031-2-A-6, pt. 2, J.P. L'Allier, "L'importance de La Téléinformatique au Québec," Allocution de M. J.P. L'Allier, Ministre des communications Prononceé par M. Florian Rompre, Sous-Ministre des Communications devant les membres de la Section de

Québec de L'Association Canadienne de l'Informatique a Québec, 24 septembre 1973.

52 SA, Gordon Grant Papers, Coll. 45, file 153, "Notes for Use of Mr. Grant in Opening Meeting of the Saskatchewan Government Computer Centre," 23 September 1968.

53 Saskatchewan Computer Utility Corporation, *Annual Report*, 1974.

54 OA, RG 14, MTC, DSCF, box T-209, file: Prairie Provinces Communications Group, "A Western Perspective on Communications," 19 November 1972.

55 NA, RG 55, Acc. 1986–87/379, box 32, file 7148-02, pt. 1, Cabinet Memorandum, "Appendix 'A,' A Discussion of Computer/Communications Policy," January 1973, quoting F.G. Withington, of Arthur D. Little Inc. in a presentation to the 1971 annual conference of the Association of Computing Machinery.

56 NA, RG 97, vol. 407, file 7500-1031-2-A-4, pt. 1, H.H. Brune, "The Feasibility of the National Computer Utility," 9 November 1973.

57 Ibid., vol. 408, file 7500-1031-2-A-6, pt. 3, Secretary of State, "Computer/Communications: Public Participation," discussion paper, 25 February 1974.

58 NA, RG 97, vol. 408, file 7500-1031-2-A-6, pt. 2, ITF/CC, Working Group on Social Aspects, "Report of the Sub-Working Group on Priorities," 25 January 1974. This was proposed before the Federal-Provincial Conference on Communications in November 1973, and fell on receptive ears at the Department of Communications. See Thomas Weissman, "Regional Policy on Computers," *Financial Post*, 8 December 1973.

59 NA, RG 97, Acc. 1990–91/170, box 1, file 5000-8, pt. 2, "Position Statement by the Government of Canada Concerning Computer/Communications Policy," 1974.

60 Ibid., vol. 403, file 7500-1031-1-2, Briefing Notes for Second Federal/Provincial Conference: Agenda Item 6.4 Computer/Communications, 1 April 1975.

61 Ibid., Acc. 1989–90/185, box 13, file 5635-1, pt. 1, George Fierheller, "Brief to the Royal Commission on Corporate Concentration," September 1975.

62 Ibid., Max Yalden to G.A. Fierheller, President SDL in Ottawa, 27 October 1975. See also T.W. Hobbs to ADMR, D.F. Parkhill, RE: Industrial Strategy Computer/Communications, 22 October 1975, D.F. Parkhill, "The Importance of the Computer/Communications Industry with Particular Reference to the Need for Canadian Control," 25 January 1974.

63 OA, RG 14, MTC Div. Subj. Correspondence files, box 06, file EDP 4141-1, D. Ham to R. Bulger Re: Computer Communications, 10 March 1976.

64 Ibid., box 7, file T.A. Croil Associates, 5005–25, "Computer/Communications Study," 1 June 1976.

Bibliography

PRIMARY SOURCES

Archives Nationales du Québec
    Côte de Fonds de la Ministère des Communications, E 10
    Côte de Fonds de la Régie des Services Publics, E 35
    Côte de Fonds du Conseil du Trésor, E 5
Archives Municipales de Montréal
    Série Conseil, Rapports et Documents
Archives of Ontario
    Records of the Ministry of Transportation and Communications, RG 14
    Records of the Ministry of Agriculture and Food, RG 16
Bell Canada Archives, Montreal
Bell Canada Regulatory Information Bank, Hull, Quebec
Canadian Pacific Archives, Montreal, Quebec
Canadian Radio-Television and Telecommunications Commission Archives
    Canadian Transport Commission Transcripts, Correspondence and Exhibits
City of Toronto Archives
    Board of Control Minutebook
    City Council Minutes
John George Diefenbaker Centre Archives, Saskatoon, Saskatchewan
    John George Diefenbaker Papers
National Archives of Canada
    Government Archives Division
    Records of the Canadian Radio-Television and Telecommunications Commission, RG 100
    Records of the Canadian Transport Commission, RG 46

Records of the Corporation Branch, RG 95
Records of the Department of National Defence, RG 24
Records of the Department of External Affairs, RG 25
Records of the Department of Industry, RG 20
Records of the Department of Finance, RG 19
Records of the Department of Consumer and Corporate Affairs, RG 103
Records of the Department of Communications, RG 97
Records of the Economic Council of Canada, RG 75
Records of the Ministry of State for Science and Technology, RG 102
Records of Parliament, RG 14
Records of the Privy Council Office, RG 2
Records of the Royal Commission on Government Organization, RG 33/46
Records of the Treasury Board, RG 55
Manuscript Division
A. Dunn Papers, MG 31, J6
Eric Kierans Papers, MG 32, B 10
John G. Diefenbaker Papers, MG 26, M
Louis St Laurent Papers, MG 26, L
Records of the Canadian Information Processing Society, MG 28, I 80
Records of the Canadian Marconi Company, MG 28, III 72
National Transportation Agency Archives
  Board of Transport Commissioners, Transcripts and Exhibits
Privy Council Office
  Cabinet Documents and Memoranda
Public Record Office, United Kingdom
  Records of the Dominions Office, DO 35
Saskatchewan Archives Board
  Gordon Grant Papers, Collection 45
University of Waterloo Archives

PUBLISHED GOVERNMENT DOCUMENTS

Board of Transport Commissioners for Canada. *Canadian Railway and Transport Cases*, annual
– *Judgements, Orders, Rulings and Regulations*, annual
– *Report of the Board of Transport Commissioners for Canada*. Ottawa: The Board, 1940–1966
– *The Board of Transport Commissioners for Canada: A Review of its Constitution, Jurisdiction and Practice*, by Rod Kerr. Ottawa: Queen's Printer and Controller of Stationery, 1957
Canada. Canadian Computer/Communications Task Force. *Branching Out: Report of the Canadian Computer/Communications Task Force*. 2 Vol. Ottawa: Department of Communications, 1972

Canada. Canadian Consumer Council. *Annual Report – Canadian Consumer Council.* Ottawa: The Council, 1969–1972
– *Report on the Consumer Interest in Regulatory Boards and Agencies.* Ottawa: The Council, 1973
Canada. Canadian Overseas Telecommunications Corporation. *Annual Report – Canadian Overseas Telecommunications Corporation,* 1951–1975
Canada. Canadian Radio-Television and Telecommunications Commission. *Annual Report.* Ottawa: The Commission, 1976–1980
Canada. Canadian Transport Commission. *Annual Report.* Ottawa: The Commission, 1967–1977
– *Transport Cases.* Ottawa: Information Canada, 1975–1977
Canada. Department of Communications. *Annual Report,* 1969–1977
– *Canadian Telecommunications: An Overview of the Canadian Telecommunications Carriage Industry.* Ottawa: National Telecommunications Branch, 1983
– *Communications: Some Federal Proposals.* Ottawa: The Department, 1975
– *Computer/Communications Policy: A Position Statement by the Government of Canada.* Ottawa: The Department, 1973
– *Press Releases, Statements, Etc.* Ottawa: Department of Communications, 1970
– *Proposals for a Communications Policy for Canada: A Position Paper of the Government of Canada.* Ottawa: Information Canada, 1973
– *Telecommission Studies* (43 Studies). Ottawa: The Department of Communications, 1971
– Telecommission Directing Committee. *Instant World: A Report on Telecommunications in Canada.* Ottawa: Information Canada, 1971
Canada. Department of Industry. *White Paper on a Domestic Satellite Communication System for Canada.* Ottawa: Queen's Printer, 1968
Canada. Dominion Bureau of Statistics. *The Consumer Price Index for Canada (1949 = 100): (revision based on 1957 expenditures).* Ottawa: Dominion Bureau of Statistics, 1961
– *Historical Catalogue of Dominion Bureau of Statistics Publications, 1918–1980.* Ottawa: DBS Library and Canada Year Book Division, 1986
Canada. Parliament. House of Commons. *Debates,* 1940–1977
– *Journals,* 1940–1977
– Standing Committee on Broadcasting, Films and Assistance to the Arts. *Minutes of Proceedings and Evidence,* 1960–1977
– Standing Committee on Railways, Canals and Telegraph Lines. *Minutes of Proceedings and Evidence,* 1945–1964
– Standing Committee on Transport and Communications *Minutes of Proceedings and Evidence,* 1965–1977
Canada. Statistics Canada. *Canadian Economic Observer: Historical Statistical Supplement.* Ottawa: Statistics Canada, 1986
– *Communications Equipment Manufacturers.* Ottawa: Statistics Canada, Manufacturing and Primary Industries Division, 1960–1977

– *Computer Service Industry.* Ottawa: Statistics Canada, Merchandising and Service Division, 1972–1977
– *Corporation Financial Statistics.* Ottawa: Statistics Canada, Business Finance Division, 1965–1977
– *Industrial Commodity Classification Manual, Volume I: The Classification: A Working Manual.* Ottawa: Statistics Canada, Manufacturing and Primary Industries Division (Information Canada), 1973
– *National Income and Expenditure Accounts Annual Estimates, 1926–1986.* [computer file]. Ottawa: Statistics Canada, Electronic Data Dissemination Division, 1988
– *Office and Store Machinery Manufacturers.* Ottawa: Statistics Canada, Manufacturing and Primary Industries Division, 1960–1977
– *Products Shipped by Canadian Manufacturers.* Ottawa: Statistics Canada, Manufacturing and Primary Industries Division, 1961–1977
– *Service Bulletin: Communications.* Ottawa: Statistics Canada, Transportation and Communications Division, 1971–1975
– *StatCan CANSIM Disc* [computer file]. Toronto: Micromedia, 1991
– *System of National Accounts: National Income and Expenditure Accounts.* Ottawa: Statistics Canada, Gross National Product Division, 1953–1968
– *Telephone Statistics: Preliminary Report on Large Telephone Systems.* Ottawa: Statistics Canada, Communications Section, 1955–
– *Telephone Statistics,* annual. Ottawa: Statistics Canada, Transportation and Communications Division, Communications Section, 1945–1975
– *Trade of Canada: Imports by Countries.* Ottawa: Statistics Canada, External Trade Division, 1944–1977
Commonwealth Telecommunications Board. *Annual Report and Statement of Accounts,* 1950–1967
Consultative Committee on the Implications of Telecommunications for Canadian Sovereignty. *Telecommunications and Canada / Consultative Committee on the Implications of Telecommunications for Canadian Sovereignty.* Ottawa: The Committee, 1979
Federal-Provincial Conference of Ministers Responsible for Communications. *Documents, 1973–1975.* Ottawa: Department of Communications, 1975
International Telecommunication Union. *Yearbook of Common Carrier Telecommunication Statistics 9ᵗʰ Edition (Chronological Series 1971–1980.* Geneva: ITU, 1981.
Interprovincial Conference of Communications Ministers. *Final Text of the Interprovincial Conference on Communications Held on November 21, 1972 at Quebec City, Quebec.* Quebec: Interprovincial Conference on Communications, 1972
McMullin Publishers Limited. *McGoldrick's Handbook of the Canadian Customs Tariff and Excise Duties,* annual. Montreal, 1950–1975
Ontario. Legislature of Ontario. *Debates,* 1944–1975
– Ontario Hydro-Electric Power Commission. Rural Telephone Committee. *Report to the Hydro-Electric Power Commission of Ontario Concerning Rural Telephone Service in Ontario.* Toronto: The Commission, 1953

- Ontario Railway and Municipal Board. *Annual Reports*, 1940–1953
- Ontario Telephone Service Commission. *Annual Report of the Ontario Telephone Service Commission Including Summary of Statistical Returns from Telephone Systems.* Downsview: The Commission, 1960–1976
- Ontario Telephone Authority, Department of Agriculture. *The Ontario Telephone Authority.* Toronto: The Authority, 1954

Québec. Assemblée Nationale. Commission Permanente de L'Education, des Affaires Culturelles et des Communications. *Journal des Débats*, 1967–1976
- Assemblée Nationale. *Journal des Débats.* Québec: Assemblée Nationale, 1944–1977
- Ministère des Communications de Québec. *Québec: Maître d'Œuvre de la Communication sur son Territoire.* Québec: MCQ, 1971
- Ministère des Communications de Québec. *Rapport des Activités*, 1969–70 to 1975–76
- Régie des Services Publics, *Rapport Annuel.* Québec: La Régie, 1965–1977
- Ministère des Communications de Québec. Comité Interministeriel sur la Téléinformatique. *Rapport Intérimaire.* Québec: MCQ, 1973
- Régie des Services Publics. *La Régie des Services Publics du Québec et le Contrôle des Services Téléphoniques* par John D. Gregory. Québec: Ministère des Communications, 1975

Saskatchewan. Department of Telephones. *Annual Report.* Regina, Saskatchewan: Department of Telephones, 1944–1977
- Legislative Assembly. *Debates and Proceedings.* Regina: Legislative Assembly, 1948–1975
- Saskatchewan Computer Utility Corporation. *Annual Reports*, 1973–1977

SECONDARY SOURCES

Anderson, Douglas. *Regulatory Politics and Electric Utilities: A Case Study in Political Economy.* Boston: Auburn Publishing Co., 1981
Andrew, Edward and Zravdko Planinc. "Technology and Justice: A Round Table Discussion." Peter C. Emberley, ed. *By Loving Our Own: George Grant and the Legacy of Lament for a Nation.* Ottawa: Carleton University Press, 1990
Armstrong, Christopher. *The Politics of Federalism: Ontario's Relations with the Federal Government, 1867–1942.* Toronto: University of Toronto Press, 1981
Arrow, Kenneth J. "Economic Welfare and the Allocation of Resources for Invention." Universities-National Bureau Committee for Economic Research, Special Conference Series, *The Rate and Direction of Inventive Activity: Economic and Social Factors.* Princeton: Princeton University Press, 1962
Ashley, C.A. and R.G.H. Smails. *Canadian Crown Corporations: Some Aspects of their Administration and Control.* Toronto: Macmillan, 1965
Aufderheide, Patricia. "Universal Service Telephone Policy in the Public Interest." *Journal of Communication* 37 (1987): 81–96

Babe, Robert E. "Vertical Integration and Productivity: Canadian Telecommunications." *Journal of Economic Issues* 15 (1981): 1–31
– *Telecommunications in Canada: Technology, Industry and Government.* Toronto: University of Toronto Press, 1990
– "Control of Telephones: The Canadian Experience." *Canadian Journal of Communication* 13 (1988): 16–29
Baird, H.E. "Justifying Electronic Data Processing in Government Service." Canadian Conference for Computing and Data Processing. *Proceedings of the First Conference, University of Toronto, June 9 and 10, 1958*
Baldwin, John. *The Regulatory Agency and the Public Corporation: The Canadian Air Transport Industry.* Cambridge, MA: Ballinger, 1975
Barbe, Raoul. *Les Organismes Québécois de Régulation des Entreprises D'Utilité Publique.* Montréal: Wilson et Lafleur, 1980
Barmash, Isadore. *The World is Full of It: how we are oversold, overinfluenced and overwhelmed by the communications manipulators.* New York: Delacorte Press, 1974
Baughcum, Alan, and Gerald R. Faulhaber, eds. *Telecommunications Access and Public Policy: Proceedings of the Workshop on Local Access, September 1982, St. Louis, Missouri.* Norwood, NJ: Ablex Publishing Corporation, 1984
Baxter, S.D. "Computers and the National Research Council." *Proceedings of the Royal Society of Canada, 4th Series* 3, 1965.
Bell, Daniel. *The Coming of Post-Industrial Society: A Venture in Social Forecasting.* New York: Basic Books, 1973
Beniger, James R. *The Control Revolution. The Technological and Economic Origins of the Information Society.* Cambridge, MA: Harvard University Press, 1986
Bhaneja, B., et al. *Technology Transfer by DOC: A Study of Eight Innovations. Background Paper 12.* Ottawa: Ministry of State for Science and Technology, 1980
Bickers, Kenneth. "The Politics of Regulatory Design: Telecommunications Regulation in Historical and Theoretical Perspective." Ph.D. thesis, University of Wisconsin-Madison, 1988
Bolter, J. David. *Turing's Man: Western Culture in the Computer Age.* Chapel Hill: University of North Carolina Press, 1984
Boon, John. "Telecommunications and the Constitution." *Saskatchewan Law Review* 49 (1985): 69–88
Braeutigam, Ronald and Bruce Owen. *The Regulation Game: Strategic Use of the Administrative Process.* Cambridge, MA: Ballinger, 1978
Brauetigam, Ronald. *The Regulation of Multiproduct Firms: Decisions on Entry and Rate Structure.* Stanford, CA: Stanford University Press, 1976
Breton, A. and A.D. Scott. *The Economic Constitution of Federal States.* Toronto: University of Toronto Press, 1978
Britnell, G.E. "Public Ownership of Telephones in the Prairies." MA thesis, University of Toronto, 1934

Britton, John N.H. and Gilmour, James M. *The Weakest Link: A Technological Perspective on Canadian Industrial Underdevelopment.* Ottawa: Science Council of Canada, 1978

Brock, Gerald W. *The Telecommunications Industry: The Dynamics of Market Structure.* Cambridge, MA: Harvard University Press, 1981

Brothers, James. "Telesat Canada: Pegasus or Trojan Horse?" MA thesis, Carleton University, 1979

Brown-John, Lloyd. *Canadian Regulatory Agencies: Quis Custodiet Ipsos Custodes?* Toronto: Butterworth & Company, 1981

Brownlee, Thomas. "The Role of Communications in Economic Development," MA thesis, McGill University, 1984

– "Teleglobe Canada: Outside the (CRTC) Regulatory Camp." *Canadian Public Administration* 29 (1986): 425–44

Buchan, Robert J. and C. Christopher Johnston. "Telecommunications Regulation and the Constitution: A Lawyer's Perspective." R.J. Buchan et al., *Telecommunications Regulation and the Constitution.* Montreal: Institute for Research on Public Policy, 1982

Cadbury, George W. *Public Enterprises in the Province of Saskatchewan.* Brussels: International Institute of Administrative Sciences, 1955

Cairns, Alan C. "The Governments and Societies of Canadian Federalism." *Canadian Journal of Political Science* 10 (1977): 695–725

Campell-Kelly, Michael. *ICL: A Business and Technical History.* Oxford: Clarendon Press, 1989

Canadian Labour Congress. *Submission by the Canadian Labour Congress to the Telecommunications Committee of the Canadian Transport Commission, January 31, 1973*

Canadian Information Processing Society. *Canadian Computer Census, 1970–75.* Toronto: CIPS, various years

Canadian Marconi Company. *Submission to the Royal Commission on Canada's Economic Prospects.* Montreal: The Company, 1956

Carey, James W. and John Quirk. "The Mythos of the Electronic Revolution." *American Scholar* 39 (1970): 219–41; 395–424

Chandler, Alfred, et al. "Some Issues of Technology, Proceedings of a Conference." From a conference held in Cambridge, Massachusetts, April 30, 1979. *Daedalus* 109 (1980): 1–24

Chandler, Harry S. "The Introduction of New Technology into Telecommunications by Competition or Cooperation: The Case of Satellite Communications." MA thesis in Economics, McGill University, 1981

Chandler, William M. and Christian W. Zollner, eds. *Challenges to Federalism: Policy-Making in Canada and the Federal Republic of Germany.* Kingston: Institute of Intergovernmental Relations, Queen's University, 1989

Chisman, Forrest. "Beyond Deregulation: Communications Policy and Economic Growth." *Journal of Communication* 32 (1982): 69–83

Codding, George A. Jr and Anthony M. Rutkowski. *The International Telecommunications Union in a Changing World*. Dedham, MA: Artech House, 1982

Collins, Robert. *A Voice from Afar: The History of Telecommunications in Canada*. Toronto: McGraw-Hill Ryerson, 1977

Computing and Data Processing Society of Canada. *Proceedings of the First Conference*. Toronto: The Society, 1961

Cortada, James W. *Before the Computer: IBM, NCR, Burroughs and Remington Rand and the Industry They Created, 1865–1956*. Princeton: Princeton University Press, 1993

Costa, Regina Marie. "Past Problems and Current Issues in Canadian Telecommunication Policy: A Sense of Déja Vu." MA thesis, Simon Fraser University, 1988

Courville, Leon, Alain De Fontenay, and Rodney Dobell. *Economic Analysis of Telecommunications: Theory and Applications*. Amsterdam: North Holland, 1983

Courville, Leon and Marcel G. Digeneous. "On New Approaches to the Regulation of Bell Canada." *Canadian Public Policy* 3: (1977): 76–89

Cox, Mark. "The Transformation of Regulation: Private Property and the Problem of Government Control in Canada, 1919–1939." Ph.D. dissertation, York University, 1992

Crandall, R.W. and Kenneth Flamm, eds. *Changing the Rules: Technological Change, International Competition and Regulation in Communications*. Washington, DC: The Brookings Institution, 1989

Creighton, Donald G. *The Forked Road: Canada, 1939–1957*. Toronto: McClelland and Stewart, 1976

Cruikshank, Ken. *Close Ties: Railways, Government and the Board of Railway Commissioners, 1851–1933*. Montreal: McGill-Queen's University Press, 1991

– "The Limits of Regulation: Railway Freight Rate Regulation and the Board of Railway Commissioners, 1851–1933." Ph.D. dissertation, York University, 1988

Cuff, R.D. and J.L. Granatstein. *Ties That Bind: Canadian-American Relations in Wartime. From the Great War to the Cold War*, 2nd ed. Toronto: Hakkert, 1977

Cuff, R.D. "The State and the Growth of Economic Knowledge: A Commentary." Paper presented at the Woodrow Wilson Center Symposium, 16 September 1988

Cundiff, W.E. and Mado Reid, eds. *Issues in Canadian/US Transborder Computer Data Flows*. Proceedings of a Conference held in Montreal sponsored by the Institute for Research on Public Policy, 6 September 1978. Toronto: Butterworth and Co. (Canada) Ltd., 1979

Currie, Archibald W. "Telephone Rates in Canada." Robert M. Clark, ed. *Canadian Issues: Essays in Honour of Henry F. Angus*. Toronto: University of Toronto Press, 1961

Dalfen, C.M. "The Telesat Canada Domestic Communications Satellite." *Canadian Communications Law Review* 1 (1969): 182–211

Daly, D.J. and Steven Globerman. *Tariff and Science Policies: Applications of a Model of Nationalism*. Toronto: Ontario Economic Council, University of Toronto Press, 1976

Dameron, Kenneth. "The Consumer Movement." *Harvard Business Review* 18 (1939): 271–89

David, E.E. Jr. "Computing from the Communications Point of View." Franz L. Alt and Morris Rubinoff, eds. *Advances in Computer Systems, Vol. X*. London and New York: Academic Press, 1970

de Sola Pool, Ithiel. *Technologies without Boundaries: On Telecommunications in a Global Age*. Cambridge, MA: Harvard University Press, 1990

– "The Rise of Communications Policy Research." *Journal of Communication* 24 (1974): 31–42

Derrick, M.B. "The Saskatchewan Medical Care Insurance Plan and the Computer." Computing and Data Processing Society of Canada Annual Conference, Banff, Alberta, 1966, *Proceedings of the Conference*

Dewalt, Bryan. *Building a Digital Network: Data Communications and Digital Telephony, 1950–1990*. Ottawa: National Museum of Science and Technology, 1992

Di Norcia, Vincent. "Communications, Time and Power: An Innisian View." *Canadian Journal of Political Science* 23 (1990): 334–57

Dobell, Rodney. "Telephone Companies in Canada: Demand, Production and Investment Decisions." *Bell Journal of Economics* 3 (1972): 175–219

Doern, G.B and James Brothers. "Telesat Canada." Allan Tupper and G.B. Doern, eds. *Public Corporations and Public Policy in Canada*. Montreal: Institute for Research in Public Policy, 1981

Dyson, Kenneth and Pete Humphreys. *The Politics of the Communication Revolution in Western Europe*. London and Totowa, NJ: North Holland, 1986

Eager, Evelyn. *Saskatchewan Government: Politics and Pragmatism*. Saskatoon: Western Producer Prairie Books, 1980

Eger, J.M. "The Global Phenomenon of Teleinformatics: An Introduction." *Cornell International Law Journal* 14 (1981): 203–26

Elixmann, Dieter and Karl-Heinz Neumann, eds. *Communication Policy in Europe*. Berlin: Wissenschaftiches Insitut fur Kommunikationsdienste GmbH, Springer-Verlag, 1989

English, H.E. *Telecommunications for Canada: An Interface of Business and Government*. Toronto: Methuen, 1973

Estabrooks, M. and R.H. Lamarche. *Telecommunications: A Strategic Perspective on Regional, Economic and Business Development. Selected Papers from a Conference held in Ottawa November 10–12, 1986*. Moncton, NB: Canadian Institute for Research on Regional Development, 1987

Faulhaber, Gerald, R. *Telecommunications in Turmoil: Technology and Public Policy*. Cambridge, MA: Ballinger Publishing Company, 1987.

Fierheller, George. *The SDL Story*. Privately published by the author, 1988

Fisher, Franklin M., James W. McKie and Richard B. Mancke. *IBM and the US Data Processing Industry: An Economic History.* New York: Praeger Special Studies, 1983

Flamm, Kenneth. *Creating the Computer: Government, Industry and High Technology.* Washington, DC: The Brookings Institution, 1987

– *Targeting the Computer: Government Support and International Competition.* Washington, DC: The Brookings Institution, 1987

Fletcher, Frederick J. and Donald C. Wallace. "Federal-Provincial Relations and the Making of Public Policy in Canada: A Review of Case Studies." Richard Simeon, res. coordinator, *Division of Powers and Public Policy.* Vol. 61 of the Studies for the Royal Commission on the Economic Union and Development Prospects for Canada. Toronto: University of Toronto Press, 1985

Fletcher, Martha, and Frederick J. Fletcher. "Communications and Confederation: Jurisdiction and Beyond." R.B. Byers and Robert W Reford, eds. *Canada Challenged: The Viability of Confederation.* Toronto: Canadian Institute of International Affairs, 1979

Fombrun, Charles, and W.G. Astley. "The Telecommunications Community: An Institutional Overview." *Journal of Communication* 32:4 (1982): 56–68

Foreman-Peck, James and Jurgen Muller, eds. *European Telecommunication Organizations.* Baden-Baden: Nomos Verlagsgesellschaft, 1988

Forester, Tom, ed. *The Information Technology Revolution.* Cambridge, MA: MIT Press, 1985

Freeman, Christopher. "The Case for Technological Determinism." Ruth Finnegan, Graeme Slaman, and Kenneth Thompson, eds. *Information Technology: Social Issues. A Reader.* London: Hodder and Stoughton in association with the Open University, 1987

Friedrichs, G. and Adam Schaff, eds. *Microelectronics and Society: For Better or for Worse. A Report to the Club of Rome.* Oxford: Pergamon Press, 1982

Fuss, Melvyn and Leonard Waverman. *The Regulation of Telecommunications in Canada.* Ottawa: Economic Council of Canada, 1981

Gaffney, John, ed. *France and Modernization.* Aldershot, Hants, England: Avebury Publishing, 1988

Gainer, Walter D. *The Canadian Telecommunications Industry: Structure and Regulation.* Ottawa: Department of Communications, 1970

Galbraith, John Kenneth. *The Affluent Society.* Boston: Houghton Mifflin, 1958

Garant, Patrice. *La Régie des Services Publics.* Laboratoire de Recherche sur la Justice Administrative. Québec: Université Laval, 1977

Garnett, Robert W. *The Telephone Enterprise. The Evolution of the Bell System's Horizontal Structure, 1876–1909.* Baltimore: Johns Hopkins University Press, 1985

Gelinas, André. "Public Enterprise and Public Interest: Independence and Accountability Summary of Discussions." André Gelinas, ed. *Public Enterprise and the Public Interest. Proceedings of an International Seminar.* Toronto: Institute of Public Administration of Canada, 1978

Gentzoglanis, Anastassios. "Vertical Integration and Monopoly Regulation: A Case Study of the Bell Canada–Northern Telecom Complex." MA in Economics, McGill University, 1981

Gerbner, George, ed. *Communications Technology and Social Policy: Understanding the New "Cultural Revolution."* New York: Wiley, 1973

Gibbins, Roger. "Models of Nationalism: A Case Study of Political Ideologies in the Canadian West." *Canadian Journal of Political Science* 10 (1977): 341–73

Globerman, Steven. *Telecommunications in Canada: An Analysis of Outlook and Trends.* Vancouver: The Fraser Institute, 1988

Goehlert, Robert U. *Policy Studies on Communication: A Selected Bibliography.* Monticello, Ill.: Vance Bibliographies, 1984

Gordon, Richard. *Reforming the Regulation of Electric Utilities: Priorities for the 1980s.* Lexington, MA: Lexington Books, 1982

Gotlieb, Allan, Charles Dalfen, and Kenneth Katz. "The Transborder Transfer of Information by Communications and Computer Systems: Issues and Approaches to Guiding Principles." *American Journal of International Law* 68 (1974): 227–57

Gracey, Don. "Public Enterprise in Canada." André Gelinas, ed., *Public Enterprise and the Public Interest. Proceedings of an International Seminar.* Toronto: Institute of Public Administration of Canada, 1978

Granatstein, J.L. *The Ottawa Men: The Civil Service Mandarins, 1935–1957.* Toronto: Oxford University Press, 1982

– *Canada 1957–1967: The Years of Uncertainty and Innovation.* Toronto: McClelland and Stewart, 1986

– "The Road to Bretton Woods: International Monetary Policy and the Public Servant." *Journal of Canadian Studies* 16 (1981): 174–87

Grant, George. *Technology and Empire: Perspectives on North America.* Toronto: House of Anansi Press, 1969

Grant, Peter S. *Canadian Communications Law and Policy, Volume 1: Statutes Treaties, and Judicial Decisions.* The Law Society of Upper Canada, Department of Education, Toronto, Canada, 1988

Gregg, Donna. "Capitalizing on National Self-Interest: The Management of International Telecommunication Conflict by the International Telecommunications Union." *Law and Contemporary Problems* 45 (1982): 37–52

Gregory, J. "Telephone Regulation in Quebec: A Study of the Quebec Public Service Board." *Canadian Communications Law Review* 5 (1973): 1–124

Grindlay, Thomas. *The Independent Telephone Industry in Ontario: A History.* Toronto: The Ontario Telephone Service Commission, 1975

Habermas, Jürgen. *The Structural Transformation of the Public Sphere: An Inquiry into a Category of Bourgeois Society.* Trans. by Thomas Burger. Cambridge: MIT Press, 1989

Hall, Pamela, and Barry Lesser. *Telecommunications Services and Regional Development: The Case of Atlantic Canada.* Halifax: The Institute for Research on Public Policy, 1987

Hayes, F. Ronald. *The Chaining of Prometheus: Evolution of a Power Structure for Canadian Science*. Toronto: University of Toronto Press, 1973

Helliwell, John F. *Taxation and Investment: A Study of Capital Expenditure Decisions in Large Corporations*. Studies of the Royal Commission on Taxation No. 3. Ottawa: The Commission, 1964

Hendry, J. *Innovating For Failure: Government Policy and the Early British Computer Industry*. Cambridge, MA: MIT Press, 1989

Hermann, Robert O. "The Consumer Movement in Historical Perspective." David A. Aaker and George S. Day, eds. *Consumerism. Search for the Consumer Interest*, 3rd ed. New York: The Free Press, 1978

Hilliker, John F., ed. *Documents Rélatifs Aux Rélations Extérieures Du Canada/ Documents on Canadian External Relations*, Vols. 8–12. Ottawa: Department of External Affairs, various years

Hills, Jill. *Deregulating Telecoms: Competition and Control in the United States, Japan and Britain*. New York: Quorum Books, 1986

– "Neo-conservative Regimes and Convergence in Telecommunications Policy." *European Journal of Policy Research* 17 (1989): 95–114

Hindley, M. Patricia, Gail M. Martin, and Jean McNulty. *The Tangled Net. Basic Issues in Canadian Communications*. Foreword by George Woodcock. Vancouver: J.J. Douglas, 1977

Hockney, R.W. and C.R. Jesshope. *Parallel Computers: Architecture, Programming and Algorithms*. Bristol, UK: Adam Hilger, 1981

Hogarth, William. "A History of the Canadian Transport Commission." BA thesis, Carleton University, 1974

Hughes, Thomas P. *Networks of Power: Electrification in Western Society, 1880–1930*. Baltimore, Md: Johns Hopkins University Press, 1982

– "The Order of the Technological World." A.R. Hall and N. Smith, eds. *History of Technology, Vol. 5*. London: Mansell Publishing, 1980

– "The Dynamics of Technological Changes: Salients, Critical Problems and Industrial Revolution." Giovanni Dosi, Renato Giannetti, and Pier Angelo Toninelli, eds. *Technology and Enterprise in a Historical Perspective*. Oxford: Clarendon Press, 1992

International Telecommunications Union. *From Semaphore to Satellite*. Geneva: ITU, 1965

Irwin, Manley. "Global Cross-Currents of an Information Economy." C.C. Gotlieb, ed. *The Information Economy: Its Implications for Canada's Industrial Strategy. Proceedings of a Conference held at Erindale College, May 30–June 1, 1984*. Ottawa: Royal Society of Canada, 1985

Janisch, Hudson, ed. *Communications Law II, Volume I*. Toronto: University of Toronto Law School, 1991

Janisch, Hudson N. "The North American Telecommunications Industry from Monopoly to Competition." K. Button and D. Swann, eds. *The Age of Regulatory Reform*. Oxford: Clarendon Press, 1989

- *The Regulatory Process of the Canadian Transport Commission.* Ottawa: Law Reform Commission of Canada, 1978
Janisch, Hudson N. and M. Irwin. "Information Technology and Public Policy: Regulatory Implications for Canada." *Osgoode Hall Law Journal* 20 (1982): 610–41
Janisch, Hudson N., S.G. Rawson, and W.T. Stanbury. *Canadian Telecommunications Regulation Bibliography/Bibliographie de la reglementation des télécommunications au Canada.* Ottawa: Canadian Law Information Council, 1987
Jelly, Doris H. *Canada: 25 Years in Space.* Montreal: Polyscience Publications, 1988
Joskow, Paul L. and Richard Schmalensee. *Markets for Power: An Analysis of Electric Utility Deregulation.* Cambridge, MA: MIT Press, 1983
Kahn, Alfred E. *The Economics of Regulation. Principles and Institutions.* 2 vols. New York: John Wiley, 1970
Kaiser, Gordon. "Competition in Telecommunications: Refusal to Supply Facilities By Regulated Common Carriers." *Ottawa Law Review* 13 (1981): 95–122
Katz, Kenneth. *Regulation of Federal Data Banks: A Study for the Privacy and Computers Task Force.* Ottawa: DOC, 1973
Keefer, T.C. *Philosophy of Railroads and Other Essays.* Edited with an introduction by H.V. Nelles. Toronto: University of Toronto Press, 1972
Kemp, Kathleen. "Political Parties, Industrial Structures and Political Support For Regulation." Robert Eyestone, ed. *Public Policy Formation.* Greenwich, CT: JAI Press, 1984
Kennedy, Noah. *The Industrialisation of Intelligence: Mind and Machine in the Modern Age.* London: Unwin and Hyman, 1989
Krasnow, Erwin G., H.A. Terry, and L.D. Longley. "Rewriting the 1934 Communications Act, 1976–1980: A Case Study of the Formulation of Communications." *Comm/Ent* 5 (1981): 345–78
Lacroix, Jean-Guy. "Les Libéraux et la culture: de l'unité nationale a la marchandisation de la culture, 1963–1984." Yves Belanger, Dorval Brunelle, et al., eds., *L'ere des Libéraux: Le Pouvoir fédéral de 1963 à 1984.* Sillery, PQ: Presses de l'Université du Québec, 1988
Landes, David. *The Unbound Prometheus: Technological Change and Industrial Development in Western Europe from 1750 to the Present.* London: Cambridge University Press, 1968
Lapointe, Alain. *L'Évolution des rélations entre Bell Canada et l'organisme réglementeur: une analyse interprétative.* Ph.D. thesis, Université de Montréal, 1985
Latham, Robert. "The Telephone and Social Change." Ben D. Singer, ed. *Communications in Canadian Society.* Toronto: Addison Wesley, 1983
Latour, Bruno. *Science in Action: How to Follow Scientists and Engineers through Society.* Cambridge, MA: Harvard University Press, 1987

Lears, T.J. Jackson. "A Matter of Taste: Corporate Cultural Hegemony in a Mass-Consumption Society." Lary May, ed. *Recasting America: Culture and Politics in the Age of the Cold War.* Chicago: University of Chicago Press, 1989

Lukasiewicz, J. "A New Role for Canada: Warning Post Against Rampant Technology." *Science Forum,* February 1970

Mackay, Donald. *The People's Railway: A History of Canadian National.* Toronto: Douglas and McIntyre, 1992

Mamrak, Sandra, A. "The Design and Development of Resources – Sharing Services in Computer Communications Networks: A Survey." Morris Rubinoff and Marshall C. Yovits, eds. *Advances in Computer Systems, Vol. XVI.* London and New York: Academic Press, 1977

Mansell, Robin E. "Industrial Strategies of the Communications Information Sector: An Analysis of Contradictions in Canadian Policy and Performance." Ph.D. thesis, Simon Fraser University, 1984

Marie, G.C. "Telecommunications and Computers. Technology and the Law." *Corporate Law in the 80s: Special Lectures.* Don Mills: R. De Boo, Law Society of Upper Canada, 1982

Marshall, J.T. "The Philosophy of the Government Committee on Electronic Computers." The Computing and Data Processing Society of Canada, *Proceedings of the Third Conference.* McGill University Montreal, 1964

Martin, James. *Telecommunications and the Computer,* 2nd ed. Englewood Cliffs, NJ: Prentice-Hall, 1976

Marvin, Carolyn. *When Old Technologies Were New: Thinking about Electric Communication in the Late Nineteenth Century.* New York and Oxford: Oxford University Press, 1988

Mathews, Walter M., ed. *Monster or Messiah?: The Computer's Impact on Society.* Jackson, Mississippi: University Press of Mississippi, 1980

Mathison, Stuart L. and Philip M. Walker. *Computers and Telecommunications: Issues in Public Policy.* Englewood Cliffs, NJ: Prentice-Hall, 1970

Mattelart, Armand. "Communications in Socialist France: The Difficulty of Matching Technology with Democracy." Cary Nelson and Lawrence Grossberg, eds. *Marxism and the Interpretation of Culture.* Urbana and Chicago: University of Illinois Press, 1988

– "For a Class and Group Analysis of Popular Communication Struggles." Armand Mattelart and Seth Siegelaub, eds. *Introduction to Communication and Class Struggle: An Anthology in Two Volumes.* Vol. 2. New York: International General, 1983

– "Communications in Socialist France: the Difficulty of Matching Technology with Democracy." Cary Nelson and Lawrence Grossberg, eds. *Marxism and the Interpretation of Culture.* Urbana and Chicago: University of Illinois Press, 1988

Mattelart, Armand and Yves Stourdze. *Technology, Culture and Communication: A Report to the French Minister of Research and Industry.* Amsterdam: North-Holland Publishing, 1985

McCraw, Thomas. *Regulation in Perspective: Historical Essays.* Division of Research, Graduate School of Business Administration, Harvard University. Boston: Harvard University Press, 1981

– *Prophets of Regulation: Charles Francis Adams, James D. Brandeis, James M. Landis, Alfred E. Kahn.* Cambridge: Belknap Press of the Harvard University Press, 1984

– , ed. *Regulation in Perspective.* Cambridge, MA: Harvard University Press, 1981

McLean, J. Michael. *The Impact of the Microelectronics Industry on the Structure of the Canadian Economy.* Occasional Paper No. 8. Montreal: Institute for Research on Public Policy, March 1979

McLuhan, Marshall. *Understanding Media: The Extensions of Man.* New York: Mentor, 1964

McPhail, Thos. P. and David C. Coll. *Canadian Contributions to Telecommunications: An Overview of Significant Activities.* Calgary: University of Calgary Graduate Program in Communications Studies, 1986

McRoberts, Kenneth and Dale Posgate. *Quebec: Social Change and Political Crisis.* Toronto: McClelland and Stewart, 1977. Rev. ed. 1980.

McWhinney, Edward, ed. *The International Law of Communications.* Dobbs Ferry, NY: Oceana, 1971

Melling, J. and J. Barry. "The Problem of Culture: An Introduction." J. Melling and J. Barry, eds. *Culture in History. Production, Consumption and Values in Historical Perspective.* Exeter, UK: Exeter University Press, 1992

Mercier, Pierre Alain, Francois Plassard, Victor Scardigli. *Société Digitale: Les Nouvelles Technologies au Futur Quotidien.* Paris: Éditions du Seuil, 1984

Metropolis, N. J. Howlett, and Gian-Carlo Rota. International Research Conference on the History of Computing. *A History of Computing in the Twentieth Century – A Collection of Essays.* London: Academic Press, 1980

Meyer, J.R. et al. *The Economics of Competition in the Telecommunications Industry.* Cambridge, MA: Oelgeschlager Gunn & Hain, 1980.

Miller, J. "Policy Planning and Technocratic Power: The Significance of OTP." *The Journal of Communication* 32:1 (1982): 53–60

Mosco, V. "The Mythology of Telecommunications Deregulation." *Journal of Communication* 40:1 (1990): 36–49

Mullan, David and Roger Beaman. "The Constitutional Implications of the Regulation of Telecommunications." *Queen's Law Journal* 2 (1973): 67–92.

Mussio, Laurence B. "Prophets without Honour? Canadian Policy Makers and the First Information Highway, 1969–1975." W.T. Stanbury, ed. *Perspectives on the New Economics and Regulation of Telecommunications,* Montreal: Institute for Research on Public Policy, 1996

Nelles, H.V. and Christopher Armstrong. *Monopoly's Moment: The Organization and Regulation of Canadian Utilities 1830–1930.* Toronto: University of Toronto Press, 1988

Nelles, H.V. *The Politics of Development: Forests, Mines and Hydroelectric Power in Ontario, 1849–1941*. Toronto: Macmillan, 1974

Newman, Peter C. *Renegade in Power: The Diefenbaker Years*. Toronto: McClelland and Stewart, 1963

Nilles, Jack M. *The Telecommunications-Transportation Tradeoff. Options for Tomorrow*. New York: Wiley-Interscience Publication, 1976

Noam, Eli M. "The Public Telecommunications Network: A Concept in Transition." *Journal of Communication* 37:1 (1987): 30–48.

Northern Electric Company. *Statement on the Electrical Manufacturing Industry: Submitted to the Royal Commission on Canada's Economic Prospects*. Montreal: The Company, 1956

– *Submission to the Royal Commission on Corporate Concentration*. Mississauga: The Company, 1975

O'Connor, John Graham. "Le Pouvoir Quasi-reglementaire et L'Exercice de la Discretion Administrative en Droit Public avec Référence Spéciale à la Régulation des Télécommunications." LL.M., Université Laval, 1981

Oettinger, Anthony G. et al. *High and Low Politics: Information Resources for the 1980s*. Cambridge, MA: Ballinger Publishing Company, 1977

Ogle, E.B. *Long Distance Please: The Story of the Trans-Canada Telephone System*. Toronto: Collins, 1979

Organization for Economic Cooperation and Development. *High Level Conference on Information, Computer and Communications Policies for the 1980s*. Paris: OECD, 1980

Pandolfelli, Michele. *Regolamentazione e Deregolamentazione nelle telecomunicazioni: l'esperienza statunitense, inglese e giapponese*. Milan: Centro Italiano di Ricerca e d'Informazione, 1991

Parker, I. "Options in Telecommunications Regulation." *Canadian Journal of Communication* 15 (1990): 33–45.

Parkhill, D.F. "The Challenge of Computer Communications to Canada." *Proceedings of the Royal Society of Canada, 4th Series* 9 (1971).

Pelton Joseph. "Life in the Information Society." Jerry L. Salvaggio, ed. *Telecommunications: Issues and Choices for Society*. New York: Longman, 1983

Pepper, R.V. "Restricted Monpolies or Regulated Competitors? The Case of the Bell Operating Companies." *Journal of Communication* 27 (1987): 64–72

Phister, Montgomery. *Data Processing Technology and Economics*. Santa Monica, CA: The Santa Monica Publishing Company, 1976

Pinard, Maurice and Richard Hamilton. "The Independence Issue and the Polarization of the Electorate: The 1973 Quebec Election." *Canadian Journal of Political Science* 10 (1977): 215–59.

Porat, Marc U. "Communications Policy in an Information Society." Glen O. Robinson, ed. *Communications for Tomorrow: Policy Perspectives for the 1980s*. New York: Praeger, 1978

Powers, Eva. "Public Interest Implications of Telecommunications Deregulation." *Policy Studies Journal* 16 (1987): 146–59

Pritchard, J.R.S., ed. *Crown Corporations in Canada: The Calculus of Instrument Choice.* Toronto: Butterworths, 1983.

Purdy, H.L. *Transport Competition and Public Policy in Canada.* Vancouver: UBC Press, 1972

Pyke, Thomas N. Jr. "Time Shared Computer Systems." Franz L. Alt and Morris Rubinoff, eds. *Advances in Computer Systems, Vol VIII.* London and New York: Academic Press, 1967

Rabinovitch, Robert. "Communication Policy and Planning in Canada." Syed A. Rahim and John Middleton, eds. *Perspectives in Communication Policy and Planning.* Honolulu: East-West Center, 1977

Raboy, Marc. *Missed Opportunities: The Story of Canada's Broadcasting Policy.* Montreal: McGill-Queen's University Press, 1990

Rea, K.J. *The Prosperous Years: The Economic History of Ontario, 1939–1975.* Toronto: University of Toronto Press, 1985

Reid, Ian. "Profile of the Canadian Computer Industry." *Executive* 10 (December 1968)

Rens, Jean-Guy. *L'Empire Invisible. Histoire des Télécommunications au Canada. Volume I: De 1846 à 1956; Volume II: De 1956 à nos jours.* Sainte-Foy, QC: Presses de l'Université du Québec, 1993

Robbins, Kevin and Frank Webster. "Cybernetic Capitalism: Information Technology Everyday Life." V. Mosco and J. Wasko, eds. *The Political Economy of Information.* Madison, Wisconsin: University of Wisconsin Press, 1988.

Robinson, Gertrude J. "Prologue: Canadian Communications Studies: a Discipline in Transition?" *Canadian Journal of Communication* 12 (1987): 1–5

– "Germany's Satellite Policy Debate: Its Relevance to Europe and Canada." *Canadian Journal of Communication* 13 (1988): 23–49.

Romanow, Martin. "An Examination of the Governor-in-Council's direction-making and review powers with respect to the CRTC Authority to regulate broadcasting and telecommunications in Canada." LL.M, University of Alberta, 1984

Rose, Hilary and Steven Rose. *Science and Society.* London: Allen Lane, 1969

Rosenberg, Nathan. "Science and Technology in the Twentieth Century." Giovanni Dosi, Renato Giannetti, and Pier Angelo Toninelli, eds. *Technology and Enterprise in a Historical Perspective.* Oxford: Clarendon Press, 1992

Rositi, Franco. *Razionalità sociale e technologiche dell' informazione: descrizione e critica dell'utopia tecnocratica.* Milan: Edizioni di Communità, 1973

Rowlands, H.W. "The Canadian Scene in Computing and Data Processing." The Canadian Conference for Computing and Data Processing, *Proceedings of the First Conference.* University of Toronto, 9 and 10 June 1958

Salter, Liora and Debra Slaco. *Public Inquiries in Canada.* Science Council of Canada Background Study 47. Ottawa: Supply and Services, 1981

Schiller, Herbert I. "The Appearance of National Communications Policies: A New Arena for Social Struggle." *Gazette* 2 (1975): 000.

Schmalensee, Richard. *The Control of Natural Monopolies*. Lexington, MA: Lexington Books, 1979

Schmandt, Jürgen, Frederick Williams, Robert H. Wilson, and Sharon Strover, eds. *Telecommunications and Rural Development: A Study of Private and Public Sector Innovation*. New York: Praeger, 1991

Schultz, Richard. *Federalism, Bureaucracy and Public Policy: The Politics of Highway Transport Regulation*. Montreal: Institute of Public Administration of Canada, 1980

– "Regulation and Public Administration." *Canadian Public Administration* 25 (4) 1982

– *Federalism and the Regulatory Process*. Montreal: Institute for Research on Public Policy, 1979

Schultz, Richard and Alan Alexandroff. *Economic Regulation and the Federal System*. Toronto: University of Toronto Press, 1985

Schwartz, Louis B. "The Art of Telecommunications Regulation." *Journal of Communication* 34:4 (1984): 180–7.

Scitovsky, Tibor. *The Joyless Economy: The Psychology of Human Satisfaction and Consumer Dissatisfaction*, rev. ed. Oxford: Oxford University Press, 1992

– "On the Principle of Consumer Sovereignty" in *American Economic Review* 52, 2nd Series (1962): 262–8.

Serafini, S. and Andrieu, M. *The Information Revolution and Its Implications for Canada*. Ottawa: DOC, Communications Economics Branch, 1981

Servan Schreiber, Jean Jacques. *The American Challenge*. Trans. Ronald Steel. New York: Athenaeum, 1968

Sharp, I.P. Associates. *A Company Profile*. Toronto: I.P. Sharp Associates, 1980

Simeon, Richard. *Federal-Provincial Diplomacy: The Making of Recent Policy*. Toronto: University of Toronto Press, 1972

Skowronek, Stephen. *Building a New American State: The Expansion of National Administrative Capacities, 1877–1920*. Cambridge: Cambridge University Press, 1982

Slater, David W. *Consumption Expenditures in Canada*. Study for the Royal Commission on Canada's Economic Prospects. Ottawa: The Commission, 1957

Smith, Denis. *Gentle Patriot: A Political Biography of Walter Gordon*. Edmonton: Hurtig Publishers, 1973

Smith, J.M. *Canadian Economic Growth and Development from 1939 to 1955*. Study for the Royal Commission on Canada's Economic Prospects. Ottawa: The Commission, 1957

Smith, Michael L. "Representations of Technology at the 1964 World's Fair." Richard Wightman Fox, and T.J. Jackson Lears, eds., *The Power of Culture: Critical Essays in American History*. Chicago: University of Chicago Press, 1993

Smythe, Dallas W. "Culture, Communications Technology and Canadian Policy." *Canadian Journal of Communication* 12 (1986): 1–20.

Snow, Marcellus S., ed. *Marketplace for Telecommunications: Regulation and Deregulation in Industrialized Democracies*. White Plains, New York: Longman, 1986

– "The State as Stopgap: Social Economy and Sustainability of Monopoly in the Telecommunications Sector." *Review of Social Economy* 46 (1988): 1–23.

Staudenmaier, John M. *Technology's Storytellers: Reweaving the Human Fabric*. Boston: MIT Press, 1985

Stigler, George. *The Citizen and the State: Essays on Regulation*. Chicago: University of Chicago Press, 1975

Stone, Alan. *Public Service Liberalism: Telecommunications and Transition in Public Policy*. Princeton: Princeton University Press, 1991

Stray, J.F. *Inside an International. Forty Years in Cable and Wireless*. London: Regency Press, 1982

Surtees, Lawrence. *Pa Bell: A. Jean de Grandpré and the Meteoric Rise of Bell Canada Enterprises*. Toronto: Random House, 1992

Tardi, Gregory. "The Appointment of Federal Regulatory Commissioners: A Case Study of the CRTC." *Canadian Public Administration* 24 (1981): 587–95.

Teheranian, Majid, Farad Hakimyadeh, and Marcello L. Viadale. *Communications Policy for National Development: A Comparative Perspective*. Boston: Routledge and Kegan Paul, 1977

Terleckyj, Nestor E. "The Growth of the Telecommunications and Computer Industries." A. Crandall and K. Flamm, eds. *Changing the Rules: Technological Change, International Competition and Regulation in Communications*. Washington: The Brookings Institution, 1989

Thomas, David. *Knights of the New Technology: The Inside Story of Canada's Computer Elite*. Toronto: Key Porter, 1983

Trebilcock, M.J. *The Case for A Consumer Advocate*. Ottawa: Canadian Consumer Council, 1972

Tupper, Allan and G.B. Doern. "Public Corporations and Public Policy in Canada." Allan Tupper and G.B. Doern, eds. *Public Corporations and Public Policy in Canada*. Montreal: Institute for Research in Public Policy, 1981

Van Allen, W.H. "The New Commonwealth Communication System." *Canadian Geographical Journal* 45:4 (1962): n.p.

Vietor, Richard H.K. *Energy Policy in America since 1945. A Study of Business Government Relations*. Cambridge: Cambridge University Press, 1984

Vig, Norman J. *Science and Technology in British Politics*. London: Pergamon Press, 1968

Walker, Charles R. *Technology, Industry and Man: The Age of Acceleration*. New York: McGraw-Hill, 1968

Ward, Barbara. *Spaceship Earth*. New York: Columbia University Press, 1966

Ward, Norman and Duff Spafford, eds. *Politics in Saskatchewan*. Don Mills: Longmans Canada, 1968

Werner, Manuel. "The Transborder Data Flow Phenomenon: A Multinational Enterprise Information Flow Perspective." Ph.D. thesis, Université de Montréal, 1984

Will, Robert M. *Canadian Fiscal Policy, 1945–1963*. Studies of the Royal Commission on Taxation No. 17. Ottawa: The Commission, 1966

Williams, Frederick. *The Communications Revolution*. Beverley Hills: Sage Publications, 1982

Williams, Glen. *Not for Export: Toward a Political Economy of Canada's Arrested Industrialization*. Toronto: McClelland and Stewart, 1983

Wilson, James Q., ed. *The Politics of Regulation*. New York: Basic Books, 1980

Winner, Langdon. *Autonomous Technology: Technics out-of-control as a Theme in Political Thought*. Cambridge, MA: MIT Press, 1977

Wolfe, David. "Mercantilism, Liberalism and Keynesianism." *Canadian Journal of Political and Social Theory* 5 (1981): 69–96

Woodcock, George. *Strange Bedfellows: The State and the Arts in Canada*. Vancouver: Douglas and McIntyre, 1985

Woodrow, R Brian. "Players, Stakes and Politics in the Future of Telecommunications Policy and Regulation in Canada." W.T. Stanbury, ed. *Telecommunications Policy and Regulation*. Montreal: Institute for Research on Public Policy, 1986

Woodrow, R.Brian, Kenneth Woodside, Henry Wiseman, and J.B. Black. *Conflict over Communications Policy: A Study of Federal-Provincial Relations and Public Policy*. Montreal: C.D. Howe Institute, 1980

Young, R.A., Phillippe Faucher, and André Blais. "The Concept of Province-Building: A Critique." *Canadian Journal of Political Science* 17 (1984): 783–818

Zysman, John. *Political Strategies for Industrial Order. State, Market and Industry in France*. Berkeley: University of California Press, 1977.

Index

Action Bell Canada (consumer group), 134

Agricultural Rehabilitation and Development Agency (ARDA), 85

Alouette I, 97

Alouette II, 97

American Telephone and Telegraph (AT&T): conflict with Canadian cabinet, 59; relationship with COTC, 58–9

Anzac Scheme, 51

Archibald, W., chief commissioner BTC, 22

Association of Communications Regulatory Bodies, 186

Association of Municipalities of Ontario. *See* Bell CTC Rate Hearing "B" 1974

Atkey, Ronald, 145; view of federal-provincial discussions in communications, 183–4

automatic rate adjustment formula, 147–50; failure of, 149; rationale, 147–8; as solution to regulatory problems, 148

Avro Aircraft of Malton, 74

Balcer, Léon, 112

Beaverbrook, Lord, 54

Beckett, W.A., 127

Bell Canada. *See* Bell Telephone Company of Canada

Bell BTC rate hearings
- *1949–50*, 22–5; Bell position, 22; Bell strategy, 24; final decision on application, 24; interim relief question, 22, 24; municipal opposition, 22–3; reaction to decision, 24
- *1951*, 24–6; Bell position, 24–5; Bell stock performance, 25; Canadian Federation of Mayors and Municipalities, 25; Confédération des Travailleurs Catholiques du Canada, 25; decision, 25–6; municipal opposition, 25
- *1957–8*, 29–38; appeal to cabinet, 32–3; Bell position, 29–30, 34; Bell public relations offensive, 35; cabinet action to suspend board decision, 33; decision appealed to cabinet, 37; Diefenbaker intervention, 37; effects of rescission on BTC, 34; fears of technological lag, 36; income tax ruling and implications, 31–2; judgment, 31; municipal organization and opposition, 30, 31, 36, 36; opposition to increases, 30, 34; response of organized labour, 30; share price considerations, 30; support for Bell, 35–6; Toronto Board of Trade support for Bell, 35
- *1965*: board decision, 43; main issues, 42; permitted level of earnings, 42; rate of return, 42; telephone rates, 42